Aquatic and Standing Water Plants
of the Central Midwest

Books in the Aquatic and Standing Water Plants
of the Central Midwest Series by Robert H. Mohlenbrock

Cyperaceae: Sedges

*Filicinae, Cymnospermae, and Other Monocots, Excluding Cyperaceae:
Ferns, Conifers, and Other Monocots, Excluding Sedges*

Acanthaceae to Myricaceae: Wild Petunias to Myrtles

Nelumbonaceae to Violaceae: Water Lotuses to Violets

Other Southern Illinois University Press Books
by Robert H. Mohlenbrock

Guide to the Vascular Flora of Illinois, revised and enlarged edition

Distribution of Illinois Vascular Plants, with Douglas M. Ladd

A Flora of Southern Illinois, with John W. Voigt

In the Illustrated Flora of Illinois Series

Ferns, 2nd edition

Flowering Plants: Basswoods to Spurges

Flowering Plants: Flowering Rush to Rushes

Flowering Plants: Hollies to Loasas

Flowering Plants: Lilies to Orchids

Flowering Plants: Magnolias to Pitcher Plants

Flowering Plants: Nightshades to Mistletoe

*Flowering Plants: Pokeweeds, Four-o'clocks, Carpetweeds,
Cacti, Purslanes, Goosefoots, Pigweeds, and Pinks*

Flowering Plants: Smartweeds to Hazelnuts

Flowering Plants: Willows to Mustards

Grasses: Bromus to Paspalum, 2nd edition

Grasses: Panicum to Danthonia, 2nd edition

Sedges: Carex

Sedges: Cyperus to Scleria, 2nd edition

Aquatic and Standing Water Plants of the Central Midwest

Cyperaceae

Sedges

Robert H. Mohlenbrock

Southern
Illinois
University Press

Carbondale

Cover Illlustration by Paul W. Nelson

Library of Congress Cataloging-in-Publication Data
Mohlenbrock, Robert H., 1931–
  Cyperaceae: Sedges / Robert H. Mohlenbrock.
    p. cm. — (Aquatic and standing water plants of the central Midwest)
Includes indexes.
  1. Cyperaceae—Middle West—Identification. I. Title.

QK495.C997M629 2005
584'.84—dc22
ISBN 0-8093-2628-0 (cloth : alk. paper)   2004025862

Printed on recycled paper. ♻

The paper used in this publication
meets the minimum requirements
of American National Standard
for Information Sciences—
Permanence of Paper for
Printed Library Materials,
ANSI Z39.48-1992. ⊗

This book is dedicated to Robert J. Pierce,

Charles Newling, and Thomas Heineke,

cofounders of the Wetland Training Institute,

who were responsible for getting me involved

in teaching wetland plant identification

throughout the United States.

Contents

Illustrations

Series Preface

The purpose of the four books in the Aquatic and Standing Water Plants of the Central Midwest series is to provide illustrated guides to plants of the central Midwest that may live at least three months a year in water, though a particular plant may not necessarily live in standing water during a given year. The states covered by these guides include Iowa, Illinois, Indiana, Ohio, Kansas, Kentucky, Missouri, and Nebraska, except for the Cumberland Mountain region of eastern Kentucky, which is in a different biological province. Since 1990, I have taught week-long wetland plant identification courses in all of these states on several occasions.

The most difficult task has been to decide what plants to include and what plants to exclude from these books. Three groups of plants are within the guidelines of these manuals. One group includes those aquatic plants that spend their entire life with their vegetative parts either completely submerged or at least floating on the water's surface.

This group includes obvious submerged aquatics such as *Ceratophyllum*, the Najadaceae, the Potamogetonaceae, *Elodea, Cabomba, Brasenia, Nymphaea*, some species of *Ranunculus, Utricularia*, and a few others.

Plants in a second group are called emergents. These plants typically are rooted under water, with their vegetative parts standing above the water surface. Many of these plants can live for a long period of time, even their entire life, out of the water. Included in this group are *Sagittaria, Alisma, Peltandra, Pontederia, Saururus, Justicia*, and several others.

The most difficult group of plants that I had to consider is made up of those wetland plants that live most or all of their lives out of the water, but which on occasion can live at least three months in water. I concluded that I would include within these books only those species that I personally have observed in standing water for at least three months during a year, or which have been reported in the literature as living in standing water.

In this last group, for example, I have included *Poa annua*, since Yatskievich, in his *Steyermark's Flora of Missouri* (1999), indicates that this species may occur in standing water, even though I have not observed this myself. I have included most plants of bogs, fens, and marshes if I have observed these plants actually to be in water.

It is likely that I failed to include some plants that should have been included, but that I had not observed myself.

The nomenclature I have used in these books reflects my own opinion as to what I believe the scientific names should be. If these names differ from those used by the U.S. Fish and Wildlife Service, I have indicated this. A partial list of synonymy is included for each species, particularly accounting for synonyms that have been in use for several decades.

After the description of each plant, I have indicated the habitats in which the plant may be found, followed by the states in which the plant occurs. I have indicated the U.S. Fish and Wildlife wetland designation for each species for the states that each occurs in. In 1988, the National Wetlands Inventory Section of the U.S. Fish and Wildlife Service attempted to give a wetland designation for every plant occurring in the wild in the United States. The states covered by these aquatic manuals occur in three regions of the Fish and Wildlife Service. Kentucky and Ohio are in region 1; Illinois, Indiana, Iowa, and Missouri are in region 3; and Kansas and Nebraska are in region 5. Definitions of the Fish and Wildlife Service wetland categories are:

OBL (Obligate Wetland). Occur almost always under natural conditions in wetlands, at least 99% of the time.

FACW (Facultative Wetland). Usually occur in wetlands 67%– 99% of the time), but occasionally found in non-wetlands.

FAC (Facultative). Equally likely to occur in wetlands or non-wetlands 34%–66% of the time.

FACU (Facultative Upland). Usually occur in non-wetlands 67%–99% of the time, but occasionally found in wetlands.

UPL (Upland). Occur in uplands at least 99% of the time, but under natural conditions not found in wetlands.

NI (Not Indicated). Due to insufficient information.

A plus or minus sign (+ or -) may appear after FACW, FAC, and FACU. The plus means leaning toward a wetter condition; the minus means leaning toward a drier condition.

Although the Fish and Wildlife Service made changes to the wetland status of several species in an updated version in 1997, this later list has never been approved by Congress.

Following this is one or more common names currently employed in the central Midwest. A brief discussion of distinguishing characteristics and nomenclatural notes is often included. Illustrations accompany each species, showing the diagnostic characteristics. In some of the illustrations, a gap in the stem signifies that a portion of the stem has been omitted due to space limitations.

The sequence of families in these aquatic manuals is as follows:

1. Azollaceae	16. Cyperaceae	31. Sparganiaceae
2. Blechnaceae	17. Eriocaulaceae	32. Typhaceae
3. Equisetaceae	18. Hydrocharitaceae	33. Xyridaceae
4. Isoetaceae	19. Iridaceae	34. Zannichelliaceae
5. Lycopodiaceae	20. Juncaceae	35. Acanthaceae
6. Marsileaceae	21. Juncaginaceae	36. Aceraceae
7. Onocleaceae	22. Lemnaceae	37. Amaranthaceae
8. Osmundaceae	23. Maranthaceae	38. Anacardiaceae
9. Thelypteridaceae	24. Najadaceae	39. Apiaceae
10. Pinaceae	25. Orchidaceae	40. Apocynaceae
11. Taxodiaceae	26. Poaceae	41. Aquifoliaceae
12. Acoraceae	27. Pontederiaceae	42. Asclepiadaceae
13. Alismataceae	28. Potamogetonaceae	43. Asteraceae
14. Araceae	29. Ruppiaceae	44. Balsaminaceae
15. Butomaceae	30. Scheuchzeriaceae	45. Betulaceae

46. Boraginaceae
47. Brassicaceae
48. Cabombaceae
49. Caesalpiniaceae
50. Callitrichaceae
51. Campanulaceae
52. Caprifoliaceae
53. Ceratophyllaceae
54. Cornaceae
55. Cuscutaceae
56. Elatinaceae
57. Ericaceae
58. Escalloniaceae
59. Fabaceae
60. Haloragidaceae
61. Hippuridaceae
62. Hypericaceae

63. Juglandaceae
64. Lamiaceae
65. Lentibulariaceae
66. Lythraceae
67. Malvaceae
68. Menyanthaceae
69. Myricaceae
70. Nelumbonaceae
71. Nymphaeaceae
72. Nyssaceae
73. Oleaceae
74. Onagraceae
75. Parnassiaceae
76. Plantaginaceae
77. Podostemaceae
78. Polygonaceae
79. Primulaceae

80. Ranunculaceae
81. Rhamnaceae
82. Rosaceae
83. Rubiaceae
84. Salicaceae
85. Sarraceniaceae
86. Saururaceae
87. Saxifragaceae
88. Scrophulariaceae
89. Solanaceae
90. Styracaceae
91. Ulmaceae
92. Urticaceae
93. Valerianaceae
94. Verbenaceae
95. Violaceae

This volume consists only of family 16, the Cyperaceae. A second volume will cover the remainder of the monocots, including families 1–15 and 17–34. The dicots are also in two volumes, the first covering families 35–69, the second covering families 70–95.

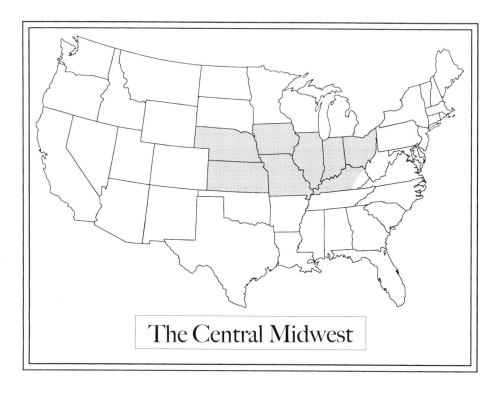

The Central Midwest

Descriptions and Illustrations

16. CYPERACEAE—SEDGE FAMILY

Annual or perennial herbs; culms frequently triangular, usually with solid intern-odes; leaves alternate, 3-ranked, less commonly absent; sheaths usually closed; ligule mostly present; inflorescence composed of 1–several spikelets variously arranged; spikelets 1- to several-flowered; flowers usually perfect (unisexual in *Carex* and *Scleria*), without a true perianth, the perianth reduced to scales, bristles, or absent; each flower subtended by a scale; stamens 1–3; ovary superior, 1-celled, with 1 ovule; stigmas 2 or 3; fruit a lenticular or trigonous achene.

This worldwide family consists of 80 to 115 genera and possibly as many as 4,000 species. Many of these are wetland and aquatic plants, and most of them have fruits useful as wildlife food.

Fifteen genera are recognized in this work as possibly having one or more species that live, at least for part of the time, in standing water.

1. Flowers bisexual; achenes not enclosed in a saclike structure (perigynium).
    2. Spikelets 1- to 2-flowered.
        3. Achene crowned with a tubercle; bristles subtending at least the lowest flower .... 12. *Rhynchospora*
        3. Achene without a tubercle; bristles absent .... 3. *Cladium*
    2. Spikelets 2- or more-flowered.
        4. Spikelets flattened; scales 2-ranked.
            5. Inflorescence terminal; bristles absent; achene without a tubercle .... 4. *Cyperus*
            5. Inflorescence axillary; bristles present; achene with a tubercle .... 5. *Dulichium*
        4. Spikelets not flattened; scales spirally arranged.
            6. Spikelet one per culm; leaf blades absent .... 6. *Eleocharis*
            6. Spikelets more than one per culm; leaf blades usually present (if absent, the inflorescence lateral).
                7. Involucral bract one, appearing like the bract of the spikelet and not appearing like the continuation of the culm .... 15. *Trichophorum*
                7. Involucral bracts 1–several, if one, then appearing like the continuation of the culm.
                    8. Flowers and fruits subtended by bristles or scales.
                        9. Flowers and fruits subtended only by scales.
                            10. Flowers and fruits subtended by minute, rudimentary scales .... 11. *Lipocarpha*
                            10. Flowers and fruits subtended by papery scales .... 10. *Hemicarpha*
                        9. Flowers and fruits subtended by capillary bristles (in addition, 3 scales present in *Fuirena*).
                            11. Flowers and fruits subtended by 3 bristles and 3 scales .... 9. *Fuirena*
                            11. Flowers and fruits subtended only by capillary bristles.
                                12. Inflorescence lateral, the culm extending beyond the inflorescence ... . 13. *Schoenoplectus*
                                12. Inflorescence terminal, the culm not extending beyond the inflorescence.
                                    13. Bristles white at maturity, forming "cottony" heads .... 7. *Eriophorum*
                                    13. Bristles reddish, brown, or tawny, not forming "cottony" heads.

14. Spikelets at least 1 cm long, the scales pubescent ....
.... 1. *Bolboschoenus*
14. Spikelets up to 1 cm long, the scales glabrous .... 14. *Scirpus*
8. Flowers and fruits not subtended by bristles or scales.
15. Inflorescence lateral, the culms extending beyond the inflorescence.
16. Achenes with vertical pebbles .... 10. *Hemicarpha*
16. Achenes cross-wrinkled .... 13. *Schoenoplectus*
15. Inflorescence terminal, the culms not extending beyond the inflorescence.
17. Leaves up to 4 mm broad .... 8. *Fimbristylis*
17. Leaves usually at least 5 mm broad .... 14. *Scirpus*
1. Flowers unisexual; achenes enclosed in a saclike structure (perigynium) .... 2. *Carex*

1. **Bolboschoenus** (Asch.) Palla—Big Bulsedge

Perennials with stout rhizomes; stems triangular in cross-section, unbranched; leaves basal and alternate, flat, glabrous, with fine, sharp, marginal teeth; ligule absent; inflorescence terminal, subtended by 2–6 bracts; spikelets 10–40 mm long, up to 12 mm across, arranged in umbels or headlike clusters; scales pubescent; spikelets many-flowered, perfect, subtended by 2–6 bristles, or the bristles sometimes absent; stamens 3; stigmas 2–3; ovary superior; achenes usually trigonous.

The species that comprise this genus have usually been assigned to *Scirpus*. When segregated as a separate genus, *Bolboshoenus* consists of 12 species found in most parts of the world. Two of these occur as aquatics or in wetlands in the central Midwest.

1. Achene sharply trigonous, 4–5 mm long; bristles subtending flowers and fruits stout, persistent .... 1. *B. fluviatilis*
1. Achene obtusely trigonous or not trigonous, 2.5–4.0 mm long; bristles subtending flowers and fruits weak, deciduous .... 2. *B. maritimus*

1. Bolboschoenus fluviatilis (Torr.) Sojak, Cas. Nar. Mus., Odd. Prir. 141:62. 1972. Fig. 1.
Scirpus maritimus L. var. *fluviatilis* Torr. Ann. Lyc. N.Y. 3:324. 1836.
Scirpus fluviatilis (Torr.) Gray, Man. Bot. 527. 1848.

Coarse perennial with moniliform rhizomes; culms stout, 3-angled, to 2 m tall; leaves scattered along the culm, alternate, flat, 8–20 mm broad; involucral leaves surpassing the inflorescence; rays of the inflorescence 6–12, to 12 cm long; spikelets 15–45 mm long, ovoid-cylindric, brown; scales ovate to ovate-lanceolate, cleft at tip, awned; bristles 6, stout, persistent, retrorsely hairy, about as long as the achene; achene obovoid, trigonous, apiculate, whitish, 4–5 mm long. May–September.

Margins of streams and lakes, marshes, sloughs, openings in bottomland forests, sinkhole ponds, often in shallow water.

IA, IL, IN, KS, KY, MO, NE, OH (OBL). The U.S. Fish and Wildlife Service calls this species *Scirpus fluviatilis*.

River bulsedge.

This is one of the most robust species of Cyperaceae. It differs from *B. maritimus* by its strongly trigonous achenes and stout bristles.

2. Bolboschoenus maritimus (L.) Palla ssp. **paludosus** (A. Nelson) A. Love & D. Love, Taxon 30 (4):845. 1981. Fig. 2.
Scirpus paludosus A. Nels. Bull. Torrey Club 26:5. 1899.

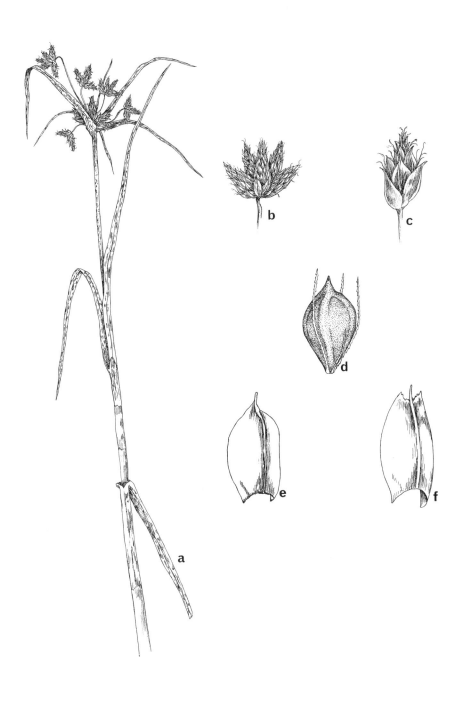

1. *Bolboschoenus fluviatilis.*
a. Habit.

b. Spikelets.
c. Spikelet.

d. Achene.
e, f. Scales.

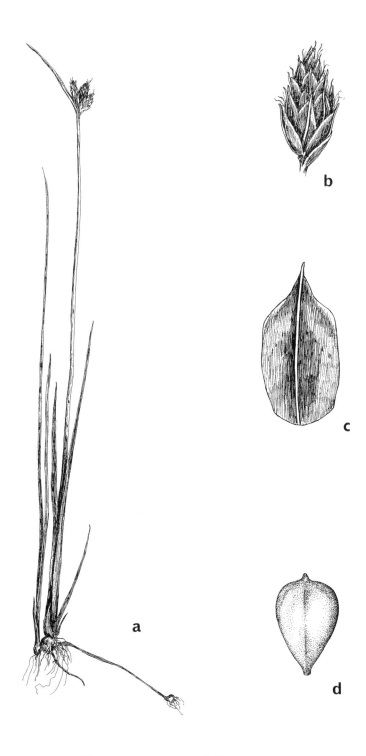

2. *Bolboschoenus maritimus* ssp. *paludosus.*   a. Habit.   c. Scale.
                                                    b. Spikelet.   d. Achene.

Scirpus maritimus L. var. *paludosus* (A. Nels.) Kuk. Repert. Spec. Nov. Regni Veg. 23:200. 1926.

Robust perennial from rather coarse rhizomes; culms to 1.5 cm tall, weakly 3-angled; leaves 3–6 in number, alternate, 5–15 mm broad, with a V-shaped, truncate ligule; involucral bracts 1–2, the longer to 20 cm long, the shorter, when present, to 10 cm long; spikelets ovoid to ovoid-cylindric, acute, 10–25 mm long, pale brown, glomerulate or on 1–4 pedicels; scales ovate, acute, pale brown; bristles weak, deciduous; achenes obovoid, apiculate, cuneate at base, weakly trigonous, 2.5–4.0 mm long, pale brown. June–October.

Alkaline wetlands, often in shallow water.

IA, IL, MO, KS, NE (NI). The U.S. Fish and Wildlife Service calls this species *Scirpus maritimus.*

Bayonet-grass.

This robust plant is sometimes considered to be a separate species from *B. maritimus,* in which case it would be called *B. paludosus*. This plant differs from *B. paludosus* by its obscurely triangular culms, its weak bristles subtending the achenes, and the smaller achenes.

2. **Carex** L.—Sedge

Perennials (in our species), often with rhizomes; culms 3-angled in cross-section; leaves 3-ranked, the lowest sometimes reduced to a bladeless sheath; blades usually flat, sometimes plicate, sometimes deeply channeled, sometimes involute, with a ligule; inflorescence of 1–many spikes, often subtended by a setaceous or leaflike bract; flowers unisexual, arranged in spikes; spikes composed of scales, each subtending a flower in its axil; staminate flowers sometimes in a separate spike, sometimes at the apex of pistillate spikes (androgynous), sometimes at the base of pistillate spikes (gynecandrous); perianth absent; stamens 1–3; ovary superior, 1-celled, with 1 ovule per cell; styles 2- or 3-cleft; achenes lenticular or trigonous, nearly entirely enclosed by a perigynium.

Carex may be comprised of as many as 2,500 species found worldwide. Although *Carex* is a major component of wetlands, nearly one-fourth of all species are upland.

1. Hairs present on leaves, sheaths, or culms (this lead does not include plant parts that are merely scabrous or perigynia that may be papillose or granular).
    2. Perigynia pubescent .... 4. *C. atherodes*
    2. Perigynia glabrous .... 29. *C. gynandra*
1. Leaves, sheaths, and culms glabrous.
    3. Perigynia pubescent.
        4. Staminate spikes 2 or more per culm.
            5. Lower spikes sometimes androgynous; pistillate scales lanceolate; pyrigynia 2.5 mm long, the beak up to 1 mm long.
                6. Culms smooth; leaves up to 2 mm broad, convolute; teeth of perigynia up to 0.6 mm long .... 39. *C. lasiocarpa*
                6. Culms usually somewhat scabrous; leaves 2–5 mm broad, flat; teeth of perigynia about 1 mm long .... 52. *C. pellita*
            5. Lowest spikes always pistillate; pistillate scales broadly ovate; perigynia 4–10 mm long, the beak 1.2–2.0 mm long .... 75. *C. trichocarpa*
        4. Staminate spike 1 per culm.

7. Leaves 2–5 mm broad; perigynia 2.0–3.5 mm long, faintly nerved .... 52. *C. pellita*
7. Leaves 4–14 mm broad; perigynia 3.2–4.5 mm long, conspicuously 2-ribbed ....
.... 60. *C. scabrata*
3. Perigynia glabrous.
  8. Stigmas 3; achenes trigonous.
    9. Spike 1, with staminate flowers at tip.
      10. Perigynia ascending, ellipsoid, 2.5–4.0 mm long .... 40. *C. leptalea*
      10. Perigynia reflexed, subulate, 6–8 mm long .... 50. *C. pauciflora*
    9. Spikes more than 1 (sometimes 1 in *C. squarrosa*), either entirely staminate or with pistillate flowers at tip.
      11. Terminal spike entirely staminate.
        12. Perigynia 10 mm long or longer.
          13. All perigynia horizontally spreading; beak of perigynium 2–3 times longer than the body; achenes truncate at summit, broader than long ...
.... 27. *C. gigantea*
          13. Perigynia ascending; beak of perigynium somewhat shorter than or slightly longer than the body; achenes narrowed at summit, longer than broad or about as long as broad.
            14. Pistillate spikes more than twice as long as broad; achenes about as long as broad, with conspicuously knobby angles and concave sides
.... 43. *C. lupuliformis*
            14. Pistillate spikes up to twice as long as broad; achenes longer than broad, without conspicuous knobby angles and with nearly flat sides
.... 44. *C. lupulina*
        12. Perigynia up to 10 mm long, usually much shorter.
          15. Staminate spikes more than 1 per culm.
            16. Plants growing in large colonies from extensive rhizomes; perigynia thick and firm, not inflated.
              17. Culms red-purple at base; leaves not glaucous or blue-green; nerves of perigynia conspicuous and slightly elevated.
                18. Teeth of the beak of the perigynium up to 1 mm long; most of the leaves 8 mm broad or broader .... 36. *C. lacustris*
                18. Teeth of the beak of the perigynium 1.0–2.2 mm long; leaves mostly 2–6 mm broad .... 37. *C. laeviconica*
              17. Culms brown at base; leaves glaucous or blue-green; nerves of perigynia faintly impressed.
                19. Nerves of perigynia faintly impressed or absent.
                  20. Nerves of perigynia faintly impressed; perigynia 6–8 mm long .... 32. *C. hyalinolepis*
                  20. Nerves of perigynia absent or nearly so; perigynia 2.8–3.5 mm long .... 28. *C. glaucescens*
                19. Nerves of perigynia coarse .... 35. *C. joori*
             16. Plants growing in dense clumps, usually without extensive rhizomes (short, stout rhizomes present in *C. vesicaria*); perigynia thin and papery, inflated.
              21. Culms spongy at base; leaves 7–12 mm broad.
                22. Pistillate scales obtuse to acute, 3–6 mm long, shorter than the perigynia .... 58. *C. rostrata*
                22. Pistillate scales acute to awned, 4–10 mm long, longer than the perigynia .... 79. *C. utriculata*
              21. Culms firm at base; leaves 2–7 mm broad.
                23. Achenes invaginated on one side; perigynia 7–10 mm long
.... 77. *C. tuckermanii*

23. Achenes not invaginated on one side; perigynia 5–8 mm long .... 80. *C. vesicaria*
15. Staminate spike 1 per culm.
    24. Perigynium prominently 2-toothed at tip, or prolonged into a conspicuous bidentate beak.
        25. Pistillate spikes 2–4 mm thick .... 81. *C. viridula*
        25. Pistillate spikes usually at least 1 cm thick.
            26. Some of the perigynia reflexed.
                27. Perigynia up to 35 per spike, 3.2–6.0 mm long, the beak about as long as the body.
                    28. Pistillate scales brownish; pistillate spikes 1.0–1.4 cm thick; perigynia yellow, 4–6 mm long, the beak 1.5–2.7 mm long .... 26. *C. flava*
                    28. Pistillate scales pale; pistillate spikes up to 1 cm thick; perigynia greenish, 3.2–4.2 mm long, the beak 1.2–1.5 mm long .... 19. *C. cryptolepis*
                27. Perigynia usually 50–100 per spike, 5–10 mm long, the beak up to 1/2 as long as the body.
                    29. Teeth of perigynium beak curved, 1.2–2.3 mm long .... 13. *C. comosa*
                    29. Teeth of perigynium beak nearly straight, 0.5–1.0 mm long .... 57. *C. pseudocyperus*
            26. All perigynia spreading to ascending, never reflexed.
                30. Culms filiform; leaves 1–3 mm broad; perigynia up to 15 per spike .... 50. *C. oligosperma*
                30. Culms broader than filiform; leaves 2.5–12.0 mm broad; perigynia (10–) 15–100 per spike.
                    31. Lowest spikes on slender, often pendulous peduncles; perigynia 4–6 mm long, 1.5–2.0 mm broad, the beak 1.8–2.2 mm long .... 34. *C. hystericina*
                    31. Lowest spikes ascending; perigynia 5–9 mm long, 2–4 mm broad, the beak 3–4 mm long.
                        32. Leaves up to 7 mm broad; perigynia 3–4 mm broad, the beak about equaling the body .... 46. *C. lurida*
                        32. Leaves 2–4 mm broad; perigynia 2.0–2.5 mm broad, the beak longer than the body ... 6. *C. baileyi*
24. Perigynium ending abruptly at tip, either without teeth or with merely a small notch.
    33. Beak or tip of perigynium bent to one side .... 72. *C. tetanica*
    33. Beak or tip of perigynium straight or nearly so.
        34. Pistillate spikes ascending to erect, not on flexuous or pendulous peduncles.
            35. Pistillate scales dark red-purple; perigynia more convex than trigonous .... 11. *C. buxbaumii*
            35. Pistillate scales not dark red-purple; perigynia trigonous.
                36. Leaves 1.5–3.5 mm wide, glaucous; pistillate scales red-brown, acute to cuspidate; perigynia ovoid, 3.0–3.5 mm long, strongly nerved .... 15. *C. crawei*
                36. Leaves 3–7 mm wide, not glaucous; pistillate scales not red-brown, awned; perigynia oblongoid, 3–4 mm long, finely impressed nerved .... 14. *C. conoidea*
        34. At least the lowest pistillate spikes on flexuous or pendulous peduncles.
            37. Perigynia glaucous-green; plants rhizomatous or stoloniferous.
                38. Leaves up to 2.5 mm wide; pistillate scales acute .... 42. *C. limosa*

38. Leaves 3–7 mm wide; pistillate scales awned .... 29. *C. glaucescens*
37. Perigynia green; plants not rhizomatous nor stoloniferous.
   39. Pistillate spikes 4–6 cm long, up to 5 mm thick; pistillate scales shorter than or about as long as the perigynia .... 55. *C. prasina*
   39. Pistillate spikes up to 4 cm long, more than 5 mm thick; pistillate scales longer than the perigynia .... 52. *C. paupercula*
11. Terminal spike with staminate flowers either at tip or at base of spike, but not throughout.
   40. Terminal spike with staminate flowers at tip .... 41. *C. leptalea*
   40. Terminal spike with staminate flowers at base.
      41. Perigynia obconic; pistillate scales setaceous, much longer than the perigynia; spikes located about midway on plant, much surpassed by the leaves and leaflike bracts .... 27. *C. frankii*
      41. Perigynia narrowly lanceoloid to narrowly ellipsoid to ovoid to obovoid (sometimes obconic in *C. typhina*); pistillate scales ovate to lanceolate, shorter than to about as long as the perigynia; most spikes in upper part of plant or even surpassing the leaves and bracts.
         42. Pistillate spikes (or if only 1 spike) 10–22 mm thick.
            43. Perigynia horizontally spreading, or the lowest ones reflexed; pistillate scales acute to acuminate to cuspidate....64. *C. squarrosa*
            43. Perigynia ascending, none of them reflexed; pistillate scales usually obtuse, less commonly acute................ 79. *C. typhina*
         42. Pistillate spikes up to 10 mm thick.
            44. At least the lowest pistillate spikes pendulous or spreading on long, slender peduncles .... 56. *C. prasina*
            44. All spikes sessile, or on short, ascending peduncles.
               45. Culms red-purple at the base; pistillate scales dark red-purple; perigynia more or less biconvex, glaucous-green; leaves glaucous below ....
                    .... 11. *C. buxbaumii*
               45. Culms usually brownish at the base; pistillate scales brown, red-brown, or greenish; perigynia trigonous, yellow, olive-green, or brown; leaves not glaucous.
                  46. Pistillate spikes 6–10 mm thick; pistillate scales acute to acuminate; perigynia 3.2–6.0 mm long, yellow or greenish, the beak about equaling the body.
                     47. Pistillate scales brownish; pistillate spikes 1.0–1.4 cm thick; perigynia yellow, 4–6 mm long, the beak 1.5–2.7 mm long ....
                        .... 26. *C. flava*
                     47. Pistillate scales pale; pistillate spikes up to 1 cm thick; perigynia greenish, 3.2–4.2 mm long, the beak 1.2–1.5 mm long ....
                        .... 19. *C. cryptolepis*
                  46. Pistillate spikes 2–3 mm thick; pistillate scales obtuse; perigynia 2–3 mm long, green or yellow-green, the beak about one-third as long as the body .... 82. *C. viridula*
8. Stigmas 2; achenes lenticular.
   48. Some or all the spikes pedunculate; staminate flowers usually on separate spikes.
      49. Terminal spike with pistillate flowers at apex.
         50. Sheaths glabrous .... 16. *C. crinita*
         50. Sheaths hispidulous .... 30. *C. gynandra*
      49. Terminal spike entirely staminate.
         51. Pistillate scales long-awned, much longer than the perigynia .... 16. *C. crinita*
         51. Pistillate scales obtuse to acuminate, shorter than to barely longer than the perigynia.

52. Lowest leaf sheath with well-developed blades; perigynia broadest above the middle .... 3. *C. aquatilis*
52. Lowest leaf sheath bladeless; perigynia broadest at or below the middle.
   53. Perigynia with a distinct beak 0.3–1.0 mm long; pistillate scales purple-black, shorter than the perigynia .... 73. *C. torta*
   53. Perigynia beakless, or the beak up to 0.3 mm long; pistillate scales red-brown, not purplish.
      54. Ligule longer than the width of the blade, V-shaped; perigynia nerveless or faintly nerved.
         55. Pistillate scales shorter than the perigynia; pistillate spikes up to 10 cm long; perigynia more or less flat, 1.7–3.4 mm long; lowest sheath deep red or purple .... 68. *C. stricta*
         55. Pistillate scales longer than the perigynia; pistillate spikes up to 5 cm long; perigynia biconvex, 1.5–2.8 mm long; lowest sheath red-brown .... 31. *C. haydenii*
      54. Ligule very short, not forming a V; perigynia distinctly nerved .... 24. *C. emoryi*
48. All spikes sessile, essentially alike, with staminate flowers either at the apex or at the base of each spike.
   56. Culms solitary, or forming stoloniferous or rhizomatous colonies.
      57. Perigynia biconvex; sheaths tight; inflorescence usually 5–10 cm long; rhizomes, if present, short and slender.
         58. Sheaths not copper-colored at summit; perigynia 2–3 mm long, about 1 mm wide, dark brown to olive-black .... 21. *C. diandra*
         58. Sheaths copper-colored at summit; perigynia 2.5–4.0 mm long, 1.2–1.3 mm wide, stramineous to brown .... 54. *C. prairea*
      57. Perigynia plano-convex; sheaths loose and open; inflorescence up to 6 cm long; rhizomes stout and thick .... 60. *C. sartwellii*
   56. Culms cespitose, not arising singly from extensive stolons and/or rhizomes.
      59. Spikes with staminate flowers at apex.
         60. Perigynia 1–3 per spike; culms capillary .... 22. *C. disperma*
         60. Perigynia 4–many per spike; culms not capillary.
            61. Culms soft, wing-angled, easily compressed; sheaths loose and open.
               62. Pistillate scales shorter than the perigynia; perigynia yellow to brown; inflorescence up to 20 cm long, up to 6 cm thick; spikes 15–25 per inflorescence.
                  63. Perigynia 3.5–6.0 mm long, strongly nerved on the inner face, the base tapering to the beak .... 66. *C. stipata*
                  63. Perigynia 6–8 mm long, nerveless or faintly nerved on the inner face, the base abruptly enlarged below the beak .... 18. *C. crus-corvi*
               62. Pistillate scales about as long as the perigynia; perigynia greenish; inflorescence up to 6 cm long, up to 1.5 cm thick; spikes 10–15 per inflorescence .... 39. *C. laevivaginata*
            61. Culms firm, at least not conspicuously wing-angled and not easily compressed; sheaths loose or tight.
               64. Sheaths not septate-nodulose; leaves 1.0–3.1 mm broad; perigynia biconvex, 1.0–1.3 mm wide; pistillate scales acute to cuspidate.
                  65. Sheaths copper-colored at summit; perigynia 2.5–4.0 mm long, 1.2–1.3 mm wide, stramineous to brown .... 54. *C. prairea*
                  65. Sheaths not copper-colored at summit; perigynia 2–3 mm long, about 1 mm wide, dark brown to olive-black .... 21. *C. diandra*

64. Sheaths septate-nodulose; leaves 2–8 mm broad; perigynia plano-convex (biconvex in *C. decomposita*), 1.2–2.0 mm wide; pistillate scales awned or at least mucronate.

  66. Perigynia plano-convex, green to stramineous to yellow- or golden brown, not spongy at the base; inflorescence up to 10 cm long, up to 1.5 cm thick.

    67. Inflorescence up to 10 cm long, usually with 15 or more spikes; perigynia 1.0–1.5 (–1.8) mm wide, the beak 0.8–1.2 mm long, about as long as the body .... 83. *C. vulpinoidea*

    67. Inflorescence up to 7 cm long, comprised of 10–15 spikes; perigynia 1.5–3.0 mm wide, the beak up to 0.7 mm long, about 1/2 as long as the body.

      68. Perigynia 2.2–3.2 mm long, 1.4–2.4 mm wide, ovate, rounded at the base .... 8. *C. brachyglossa*

      68. Perigynia 3.0–3.5 mm long, 2.5–3.0 mm wide, broadly ovate to nearly orbicular, truncate or slightly concave at the base .... 74. *C. triangularis*

  66. Perigynia biconvex, olive-black, spongy at the base; inflorescence up to 18 cm long, up to 4 cm thick .... 20. *C. decomposita*

59. Spikes with pistillate flowers at the apex.

  69. Perigynia plano-convex, with rounded margins or with narrowly rimmed margins.

    70. Perigiynia ascending, with rounded margins.

      71. Plants weak and often reclining; spikes 1–3 (–4) per culm; perigynia 1–5 per spike, oblongoid .... 77. *C. trisperma*

      71. Plants firm or, if weak, scarcely reclining; spikes usually 4 or more per culm (rarely fewer in *C. bromoides* and *C. brunnescens*); perigynia 5 or more per spike, lanceoloid to ellipsoid to narrowly ovoid (rarely only 4 in *C. bromoides*).

        72. All spikes overlapping; perigynia 3.5–4.5 mm long; beak of the perigynium 1.25–1.50 mm long .... 9. *C. bromoides*

        72. Some or all of the spikes separated from each other; perigynia 1.8–3.0 mm long; beak of the perigynium up to 1.2 mm long.

          73. Leaves up to 2 mm wide, dark green; perigynia 5–10 per spike; ventral band of sheath red-brown dotted .... 10. *C. brunnescens*

          73. Leaves 2–4 mm wide, glaucous to pale green; perigynia 10–30 per spike; ventral band of sheath not red-brown dotted .... 12. *C. canescens*

    70. Perigynia spreading to reflexed, with narrowly rimmed margins.

      74. Terminal spike usually unisexual, either entirely staminate or entirely pistillate; perigynia castaneous to nearly black .... 65. *C. sterilis*

      74. Terminal spike gynecandrous; perigynia green to dark brown.

        75. Beak of perigynium smooth; perigynia broadest near middle .... 63. *C. seorsa*

        75. Beak of perigynium serrulate; perigynia broadest near base.

          76. Beak of perigynium conspicuously notched.

            77. Midrib of pistillate scales prominent all the way to the tip .... 23. *C. echinata*

            77. Midrib of pistillate scales stopping before reaching the tip, the tip hyaline.

              78. Perigynia broadly ovoid to suborbicular, 2.0–3.5 mm

long, 1.3–2.8 mm wide, the beak 0.5–1.2 mm long .....
.... 5. *C. atlantica*

78. Perigynia lanceoloid to ovoid, 2.5–4.5 mm long, 1–2 mm wide, the beak 1–2 mm long .... 23. *C. echinata*

76. Beak of perigynium not conspicuously notched.

79. Leaves 1–3 mm wide, firm; inner face of the perigynium faintly nerved or nerveless .... 35. *C. interior*

79. Leaves up to 1 mm wide, flaccid; inner face of the perigynium strongly nerved .... 5. *C. atlantica*

69. Perigynia more or less flattened and scalelike, with winged margins.

80. Bracts leaflike and much prolonged .... 70. *C. sychnocephala*

80. Bracts not leaflike, setaceous.

81. Perigynia 7–10 mm long, the beak 4.0–4.8 mm long; spikes 12–27 mm long .... 48. *C. muskingumensis*

81. Perigynia up to 6 mm long, the beak less than 4 mm long; spikes up to 10 mm long.

82. Perigynia up to 2 mm wide.

83. Perigynia widest at the middle.

84. Perigynia evenly winged all the way to the base.

85. Perigynia appressed or ascending.

86. Perigynia lanceolate.

87. Perigynia usually nerved only on the dorsal face; sterile tufts of leaves usually not present; perigynia 3.8–5.5 mm long .... 62. *C. scoparia*

87. Perigynia nerved on both faces; sterile tufts of leaves present; perigynia 3.0–3.6 mm long ......
.... 7. *C. bebbii*

86. Perigynia ovate to obovate.

88. Perigynia ovate; pistillate scales usually awn-tipped .... 67. *C. straminea*

88. Perigynia obovate; pistillate scales acute to acuminate, not awn-tipped ..... 43. *C. longii*

85. Perigynia spreading.

89. Wing of perigynium not reaching tip of perigynium; perigynia obovate; spikes usually tapering at tip .... 2. *C. albolutescens*

89. Wing of perigynium extending to tip of perigynium; perigynia ovate; spikes usually subglobose, rounded at tip .... 49. *C. normalis*

84. Wing of perigynium narrowed above the base.

90. Perigynia lance-ovate to ovate, usually crimped on the "shoulder", all widely spreading; spikes globose .... 17. *C. cristatella*

90. Perigiynia lanceolate, not crimped on the "shoulder", at least some of them usually ascending; spikes usually not globose.

91. All but the uppermost spikes remote from each other .... 56. *C. projecta*

91. Spikes crowded, with only 1 or 2 remote spikes, if any .... 75. *C. tribuloides*

83. Perigynia widest above or below the middle, but not at the middle.

92. Perigynia widest a little above the middle; wing of the perigynium not reaching the tip of the perigynium .... 2. *C. albolutescens*

92. Perigynia widest a little below the middle, or nearly at the base; wing of the perigynium reaching the tip of the perigynium.

93. Perigynia lanceolate, the wing diminishing above the base of the perigynium ....
.... 56. *C. projecta*

93. Perigynia ovate to orbicular, the wing extending all the way to the base of the perigynium.

94. All spikes except sometimes the lowest 1–2 crowded .... 49. *C. normalis*

94. Several of the lower spikes usually remote.
  95. Pistillate scales ovate; perigynia nerved on both faces .... 71. *C. tenera*
  95. Pistillate scales lanceolate; perigynia nerved on the dorsal face but only finely nerved on the ventral face .... 25. *C. festucacea*
82. At least some of the perigynia more than 2 mm wide.
  96. Pistillate scales aristate or short-awned.
    97. Perigynia narrowly ovate, 4.8–6.0 mm long .... 32. *C. hormathodes*
    97. Perigynia broadly ovate to suborbicular, up to 5 mm long.
      98. Perigynia ovate, 2.5–3.5 mm long, widest at the middle, evenly winged all the way to the base, nerved .... 67. *C. straminea*
      98. Perigynia rhombic to suborbicular, 4–5 mm long, widest above the middle, the wing diminishing before the base or, if extending to the base, then very broad, nerveless or faintly nerved only on the ventral face.
        99. Wing of perigynium very broad, usually extending to the base of the perigynium; perigynia finely nerved on the ventral face, rounded at the base .... 1. *C. alata*
        99. Wing of perigynium not particularly broad, diminishing before reaching the base of the perigynium; perigynia more or less nerveless, cuneate at the base .... 69. *C. suberecta*
  96. Pistillate scales obtuse to acute to acuminate.
    100. Perigynia broadest above the middle.
      101. Wing of perigynium diminishing before reaching the base of the perigynium; perigynia more or less nerveless on both faces .... 69. *C. suberecta*
      101. Wing of perigynium extending to base of the perigynium; perigynia at least finely nerved on both faces.
        102. Perigynia spreading, the wing not reaching the tip of the perigynium .. .. 2. *C. albolutescens*
        102. Perigynia appressed to ascending, the wing reaching the tip of the perigynium .... 43. *C. longii*
    100. Perigynia broadest at or below the middle.
      103. Perigynia less than 4 mm long.
        104. Perigynia broadest below the middle .... 25. *C. festucacea*
        104. Perigynia broadest at the middle.
          105. Perigynia spreading.
            106. Wing of the perigynium not reaching the tip of the perigynium .... 2. *C. albolutescens*
            106. Wing of the perigynium reaching the tip of the perigynium .. .. 49. *C. normalis*
          105. Perigynia appressed to ascending .... 43. *C. longii*
      103. Some or all of the perigynia 4 mm long or longer.
        107. Perigynia nerveless on the ventral face, orbicular .... 58. *C. reniformis*
        107. Perigynia nerved, at least finely so, on the ventral face, ovate to obovate to suborbicular.
          108. Spikes globose to subglobose.
            109. Perigynia strongly nerved, 3.0–4.5 mm long, 1.5–2.1 mm wide, ovate .... 49. *C. normalis*
            109. Perigynia faintly nerved, 4.0–5.5 mm long, 2–3 mm wide, broadly ovate to suborbicular .... 47. *C. molesta*
          108. Spikes longer than broad.
            110. Perigynia spreading, the wing not reaching the tip of the perigynium .... 2. *C. albolutescens*
            110. Perigynia appressed to ascending, the wing reaching the tip of the perigynium .... 43. *C. longii*

1. **Carex alata** Torr. & Gray, Ann. Lyc. N.Y. 3:396. 1836. Fig. 3.

Plants perennial, cespitose, from short, black, fibrillose rootstocks; culms to 1.2 m tall, triangular, scabrous on the angles beneath the inflorescence, light brown near the base, with old leaf bases persisting; sterile shoots sometimes present; leaves 3–7 per culm, ascending, 2.0–5.5 mm wide, flat, firm, deep green, scabrous along the margins; sheaths tight, green-nerved ventrally, prolonged at the yellow-brown summit; spikes 3–8 per culm, 8–16 mm long, ovoid to ellipsoid, usually somewhat clavate at base, silvery brown to green, gynecandrous, usually crowded in an inflorescence up to 4 cm long; lower bracts setaceous, scabrous, the upper bracts scalelike or even absent; pistillate scales narrowly lanceolate, acuminate or more commonly awn-tipped and scabrous, usually as long as the perigynia, the center greenish and 1- to 3-nerved, the margins white-hyaline; perigynia numerous per spike, 4–5 mm long, 2.5–4.0 mm wide, obovate to suborbicular, broadest above the middle, membranous, appressed-ascending, flat, light brown to greenish, usuallly nerveless on the outer face, faintly 3-nerved on the inner face, broadly winged nearly to the base, substipitate, the beak 0.7–0.9 mm long, serrulate, bidentate; achenes lenticular, 1.6–1.9 mm long, about 1 mm wide, yellow-brown, apiculate, stipitate; stigmas 2, short, light reddish. May.

Wet ground, rarely in standing water.

IL, IN, KY, MO, OH (OBL).

Broad-winged sedge.

This species is distinguished by its spikelets with the staminate flowers at the base of each spikelet and by its suborbicular perigynia that are 4–5 mm long and that have a broad wing.

2. **Carex albolutescens** Schw. Ann. Lyc. N.Y. 1:66. 1824. Fig. 4.

Plants perennial, densely cespitose, from short, black, fibrillose rootstocks; culms to 1 m tall, triangular, slightly scabrous on the angles, at least beneath the inflorescence, longer than the leaves, light brown to brownish black at the base, with the old leaves often remaining as stubble; leaves 3–5 per culm, ascending, 2.0–3.5 mm wide, flat, pale green, the margins scabrous toward the apex; sheaths somewhat loose, green-nerved throughout or with a hyaline ventral band, prolonged at the summit; spikes 2–8 per culm, 5–13 mm long, not strongly clavate at base, greenish to stramineous, approximate or somewhat separated but rarely moniliform, the inflorescence 1.5–4.0 mm long, gynecandrous; lowest bract setaceous, the upper bracts scalelike; pistillate scales lanceolate, acute to acuminate, flat, shorter than the perigynia, the center greenish and 3-nerved, the nerves reaching the tip of the scale, the margins silver-hyaline; perigynia many per spike, 2.6–4.5 mm long, 1.5–2.7 mm wide, obovate, widest near middle, spreading, plano-convex, green to stramineous, papery, strongly nerved on the outer face, with 4–7 raised nerves on the inner face, winged to the base but not quite to the tip, substipitate, the beak 0.5–1.0 mm long, serrulate, bidentate; achenes lenticular, 1.3–1.7 mm long, 0.7–1.0 mm wide, apiculate, stipitate, jointed with the deciduous style; stigmas 2, short, reddish. April–June.

Moist woods, wet ditches, marshes, rarely in standing water.

IA, IL, IN, MO, KY, OH (FACW).

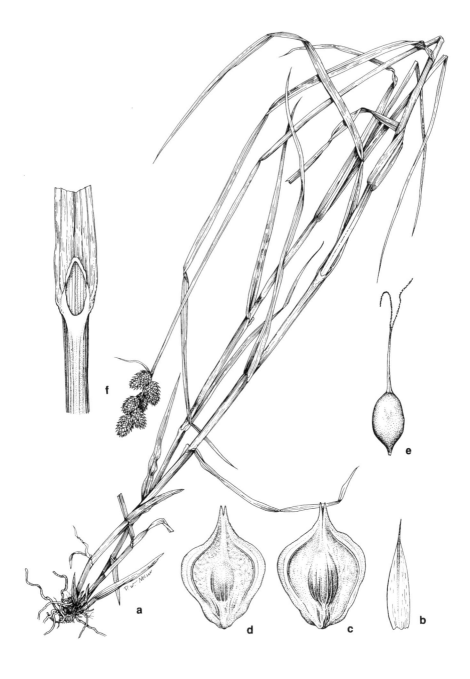

3. *Carex alata.*
a. Habit.
b. Pistillate scale.

c. Perigynium, dorsal view.
d. Perigynium, ventral view.
e. Achene.

f. Sheath with ligule.

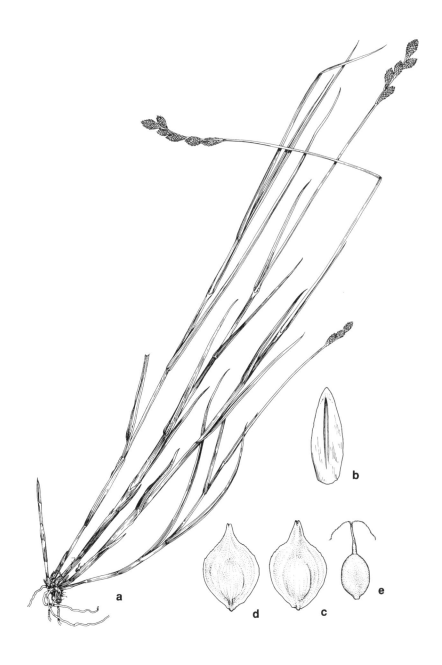

4. *Carex albolutescens.*
a. Habit.

b. Pistillate scale.
c. Perigynium, dorsal view.

d. Perigynium, ventral view.
e. Achene.

Sedge.

This species is characterized by its spikelets, which have the staminate flowers at the base, by its spreading perigynia that are winged not quite to their tip, and by the midvein of the pistillate scales reaching their tip. This species is often confused with *C. longii*, a species with ascending perigynia that are winged all the way to their tip and by their pistillate scales whose midvein does not reach their tip.

3. **Carex aquatilis** Wahlenb. var. **substricta** Kükenth. in Engl. Pflanzenr. 4 (20):309. 1909. Fig. 5.
Carex variabilis Bailey var. *altior* Rydb. Mem. N.Y. Bot. Gard. 1:76. 1900.
Carex substricta (Kükenth.) Mack. in Rydb. Fl. Rocky Mts. 139. 1938.
Carex aquatilis Wahlenb. var. *altior* (Rydb.) Fern. Rhodora 44:295. 1942.

Plants perennial, cespitose, from elongated, scaly stolons; sterile shoots common; culms to 1 m tall, smooth or scabrous on the angles, reddish brown at the base, often with last year's leaves persistent; leaves up to 15 per culm, 6.5–8.0 mm wide, flat, scabrous along the margins, papillate on both surfaces, septate between the nerves, the uppermost longer than the culms; sheaths glabrous, the hyaline ventral band usually red-speckled; uppermost 1–3 spikes staminate, up to 5 cm long, erect, densely flowered; lower 3–6 spikes pistillate, often with a few staminate flowers at the tip, erect, up to 6 cm long, densely flowered; lowest bract foliaceous; pistillate scales oblanceolate to obovate, rounded or mucronate at the tip, red-brown with a greenish midvein, a little longer to a little shorter than the perigynia; perigynia numerous, flat, 2.7–3.3 mm long, obovate, broadest above the middle, pale brown, spreading to appressed-ascending, papillate, 2-ribbed, otherwise nerveless, with a minute, entire beak, stipitate; achenes lenticular, 1.3–1.5 mm long, dark brown, short-apiculate, substipitate; stigmas 2. April–June.

Marshes, wet meadows, wet ditches, along and in streams.

IA, IL, IN, KS, MO, NE, OH (OBL).

Aquatic sedge.

If this plant is considered to be a distinct species from *C. aquatilis* var. *aquatilis*, it would be called *C. substricta*. Typical var. *aquatilis* has a less pronounced triangular stem and slightly smaller perigynia. *Carex aquatilis* var. *substricta* is similar in appearance to *C. emoryi, C. haydenii,* and *C. stricta* but differs in its well-developed blades of the lowermost sheath and in its perigynia, which are broadest above the middle.

The culms are usually reddish brown at the base, and the leaves from the previous year usually persist well into the summer.

4. **Carex atherodes** Spreng. Syst. 3:827. 1826. Fig. 6.
Carex aristata R. Br. in Richards. Frankl. Journ. 751. 1823, non Honck. (1792).
Carex trichocarpa Muhl. var. *aristata* (R. Br.) Bailey, Bot. Gaz. 10:294. 1885.

Plants perennial, cespitose, from slender, long-creeping rhizomes; culms to 1.5 m tall, triangular, usually smooth to the touch, purplish at the base; sterile shoots usually present; leaves up to 1.2 cm wide, rather thin, dull green, septate-nodulose, scabrous above, sparsely hairy beneath; sheaths often pubescent, at least near the

5. *Carex aquatilis* var. *substricta.*
a. Habit.

b. Pistillate scale.
c. Perigynium.

d. Achene.

6. *Carex atherodes.* Habit and perigynium.

summit, the ventral band often brownish, the lower sheaths purple-tinged, becoming fibrillose, the ligule longer than wide; upper 2–5 spikes staminate, to 10 cm long, to 5 mm broad, sessile except for the terminal one; staminate scales acute, awned, ciliate, yellow-brown with hyaline margins; pistillate spikes 2–4, cylindric, appressed, sometimes with a few staminate flowers at the summit, up to 12 cm long, up to 15 mm broad; pistillate scales ovate, acute, awned, reddish brown with hyaline margins and a green center, shorter than the perigynia; perigynia up to 100 per spike, lance-ovoid, 7–9 mm long, about 2 mm wide, ascending, subcoriaceous, glabrous, pale brown or yellow-green, strongly nerved, tapering to a bidentate beak with teeth 1.2–3.0 mm long; achenes trigonous, yellow-brown, substipitate; stigmas 3. May–June.

Wet meadows, marshes, occasionally in standing water.

IA, IL, IN, MO, NE, OH (OBL).

Long-beaked hairy sedge.

This robust sedge is readily distinguished by its usually pubescent leaves and sheaths, its several staminate spikes, and its strongly nerved perigynia with long-toothed beaks. It frequently grows in dense colonies.

5. **Carex atlantica** Bailey, Bull. Torrey Club 20:425. 1893.

Plants perennial, cespitose, from short rhizomes; culms triangular, scabrous, to 1 m tall; leaves 3–5 per culm, 0.5–4.0 mm wide, flat, green, scabrous along the margins, shorter than or equalling the culms; sheaths tight, smooth, the inner band hyaline and sometimes purple-dotted, cartilaginous-thickened at the concave apex, light brown; inflorescence up to 5 cm long, with 2–8 sessile, crowded or separated spikes; terminal spike 0.5–2.5 cm long, gynecandrous, the staminate part to 1.5 cm long and 2- to 20-flowered, the pistillate part to 1 cm long and 4- to 35-flowered; lateral spikes gynecandrous, 0.5–1.2 cm long, with scalelike bracts; pistillate scales ovate, 1.2–2.4 mm long, green with hyaline margins, obtuse to acute at the tip, reaching or surpassing the base of the perigynium; perigynia 2.0–3.5 mm long, 1.3–2.8 mm wide, broadly ovoid to suborbicular, plano-convex, the ventral surface with up to 12 nerves, spongy-thickened at the base, green to dark brown, sessile, tapering to a beak, with the lower perigynia spreading to reflexed; beak of the perigynium 0.5–1.2 mm long, serrulate, bidentate at the apex; achenes biconvex, 1–2 mm long, substipitate, jointed to the deciduous style; stigmas 2. April–May.

Two subspecies occur in the central Midwest, separated by the following key:

a. Larger leaves more than 1.6 mm wide; inflorescence mostly more than 2 cm long ....
.... 5a. *C. atlantica* ssp. *atlantica*
a. Larger leaves up to 1.6 mm wide; inflorescence up to 2 cm long ....
.... 5b. *C. atlantica* ssp. *capillacea*

5a. **Carex atlantica** Bailey ssp. **atlantica**. Fig. 7.
Carex incomperta Bicknell, Bull. Torrey Club 35:494. 1908.
Carex atlantica Bailey var. *incomperta* (Bicknell) F. J. Hermann, Rhodora 67:362. 1965.

Leaves more than 1.6 mm wide; inflorescence mostly more than 2 cm long.
Swampy woods, sometimes in water in depressions.

7. *Carex atlantica* ssp. *atlantica.*
a. Habit.

b. Pistillate scale.
c. Perigynium, dorsal view.

d. Perigynium, ventral view.
e. Achene.

IL, IN, MO (FACW), OH (FACW+).

Star sedge.

The perigynia are arranged in star-shaped clusters. The very narrow leaves are at least 1.6 mm wide. This subspecies is primarily a plant of the coastal plain.

5b. **Carex atlantica** Bailey ssp. **capillacea** (Bailey) Reznicek, Contr. Mich. Herb. 14:191. 1980. (Not illustrated).
Carex interior Bailey var. *capillacea* Bailey, Bull. Torrey Club 20:436. 1893.
Carex howei Mack. Bull. Torrey Club 37:245. 1910.

Leaves less than 1.6 mm wide; inflorescence up to 2 cm long.

Swampy woods, sometimes in shallow water in depressions.

IL, IN, OH (OBL).

Star sedge.

This subspecies has narrower leaves than subspecies *atlantica*. At the species level, this plant is known as *C. howei,* and that is what it is called by the U.S. Fish and Wildlife Service.

6. **Carex baileyi** Britt. Bull. Torrey Club 22:220. 185. Fig. 8.

Plants perennial, densely cespitose, from short, stout rhizomes; culms to 75 cm tall, slender, triangular, smooth or scabrous, purplish at the base; leaves 2–4 mm wide, glabrous, septate-nodulose, green, scabrous along the margins, at least some of them overtopping the culms; sheaths tight, yellowish, the ventral band hyaline, concave or truncate at the mouth, the ligule as wide as long; terminal spike 1, staminate, up to 4 cm long, up to 3 mm thick, usually short-pedunculate; staminate scales usually serrulate-awned, stramineous; lateral spikes 1–2, pistillate, slender-cylindric, 1–4 cm long, 0.8–1.3 mm thick, erect or ascending on short, smooth peduncles; bracts leaflike; pistillate scales linear, serrulate-awned, yellow-brown with hyaline margins and a green center, nearly as long as the perigynia; perigynia up to 50 per spike, crowded, 5–7 mm long, 2.0–2.5 mm thick, yellow-brown, membranous, inflated, shiny, glabrous, strongly nerved, spreading to ascending but not reflexed, abruptly tapering to a bidentate beak 3–4 mm long, the beak as long as or longer than the body; achenes trigonous, 1.5–2.0 mm long, yellow-brown, substipitate, continuous with the persistent style; stigmas 3. May–August.

Marshes, roadside ditches, rarely in standing water.

IL (not listed for region 3), KY, OH (OBL).

Bailey's sedge.

This species, whose range is generally east of the central Midwest, is similar to *C. lurida* but differs in its more slender pistillate spikes, its slightly narrower leaves, and its slightly smaller perigynia that abruptly taper into beaks that are as long as or longer than the bodies.

7. **Carex bebbii** Olney, Caric. Bor. Am. 3. 1871. Fig. 9.
Carex tribuloides Wahl. var. *bebbii* (Olney) Bailey, Mem. Torrey Club 1:55. 1889.

Plants perennial, densely cespitose, from short, compact, black to brown, fibrillose rootstocks; culms to 85 cm tall, sharply triangular, the angles scabrous beneath the inflorescence, stiff, longer than the leaves, often with sterile culms

8. *Carex baileyi.*

a. Habit.
b. Pistillate scale.
c. Perigynium.
d. Achene.

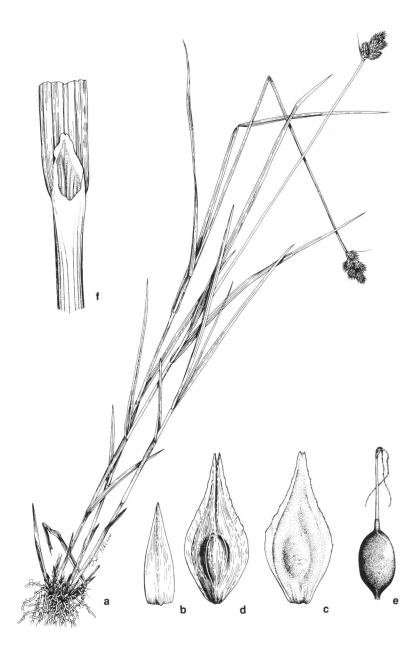

9. *Carex bebbii.*
a. Habit.
b. Pistillate scale.

c. Perigynium, dorsal view.
d. Perigynium, ventral view.
e. Achene.

f. Sheath with ligule.

present; leaves 3–5 per culm, ascending, 2–4 mm wide, the apex becoming triangular, the margins and veins scabrous, at least on the upper surface; sheaths tight, with a narrow hyaline band toward the concave summit, easily broken; spikes 3–9 per culm, ascending, lance-ovate to ovate, rounded or somewhat pointed at the tip, brownish, 5–9 mm long, gynecandrous, abruptly contracted below making the staminate flowers inconspicuous, crowded (except sometimes the lowermost) into an ovoid inflorescence up to 2 cm long; lowest bracts setaceous, scabrous, upper bracts scalelike; pistillate scales lanceolate, acute to acuminate, three-fourths to seven-eights as long as the perigynia and narrower, the center green and 3-nerved and with tan- to brown-hyaline margins; perigynia many per spike, 3.0–3.6 mm long, 1.2–1.5 mm wide, lanceolate to lance-ovate, widest at the middle, ascending to spreading, plano-convex, finely nerved only on the outer face, winged to the base, the beak 0.75–1.00 mm long, serrulate, bidentate; achenes lenticular, 1.2–1.4 mm long, 0.5–0.7 mm wide, apiculate, stipitate, light to dark brown, weakly jointed with the deciduous style; stigmas 2, short, slender, reddish brown. May–June.

Wet prairies, bogs, calcareous fens, marshes.

IA, IL, IN, OH (OBL).

Bebb's sedge.

This species is distinguished by its very narrow perigynia less than 1.5 mm wide that are nerveless on the inner face and winged all the way to the base. The crowded spikes are longer than they are broad.

8. **Carex brachyglossa** Mack. Bull. Torrey Club 50:355. 1923. Fig. 10.
Carex xanthocarpa Bicknell, Bull. Torrey Club 23:22. 1896, non Degland (1807).
Carex annectens (Bicknell) Bicknell var. *xanthocarpa* (Bicknell) Wieg. Rhodora 24:74. 1922.

Plants perennial, densely cespitose, from fibrous roots and short, stout, dark rhizomes; culms erect, up to 75 cm tall, rough to the touch beneath the inflorescence, brown at the base; leaves 3–6, up to 60 cm long, 2–5 mm wide, flat or slightly canaliculate, rough along the margins, none of them exceeding the culms; sheaths open, tight, at least the upper with the hyaline ventral band becoming septate-nodulose with age, with pale brown dots, convex at the mouth; ligule broader than long; inflorescence composed of many spikes in an elongated, interruped head up to 7 cm long and up to 1.5 cm broad; bracts setaceous, subtending most of the spikes, up to 5 cm long; spikes 10–15, the staminate flowers above the pistillate; pistillate scales lanceolate, awn-tipped, reddish brown with hyaline margins and a green midnerve, as long as or longer than the perigynia; perigynia several per spike, ovoid to ellipsoid, 2.2–3.0 mm long, 1.5–2.0 mm broad, plano-convex, yellow to golden brown except for the margins, with 3 nerves on the dorsal face, tapering to a 2-toothed, serrulate beak up to 0.7 mm long; achenes 1.2–1.5 mm long, lenticular, broadly ellipsoid, red-brown, glossy, apiculate, jointed to the style, the style enlarged at base; stigmas 2, reddish brown. May–July.

Around ponds and lakes, marshes, fens, ditches.

IA, IL, IN, KY, MO, OH (FACW), KS, NE (FAC+). The U.S. Fish and Wildlife Service calls this plant *C. annectens*, but see discussion under *C. vulpinoidea*.

Golden fox sedge.

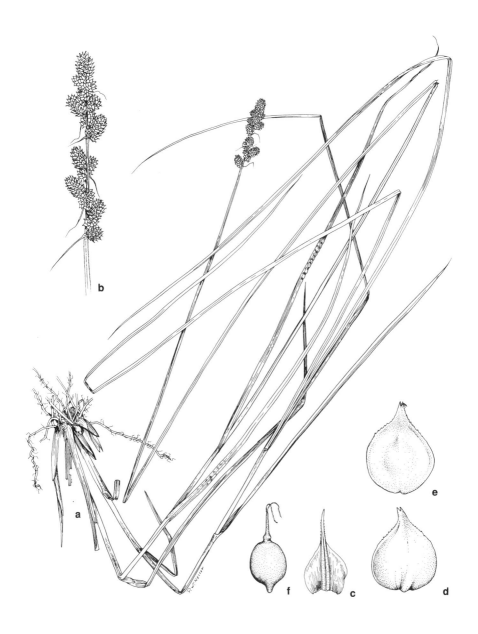

10. *Carex brachyglossa.*
a. Habit.
b. Inflorescence.
c. Pistillate scale.
d. Perigynium, dorsal view.
e. Perigynium, ventral view.
f. Achene.

This is the only species of *Carex* that has an elongated spike of 10–15 spikelets that are golden brown at maturity and the beak of the perigynium less than 0.8 mm long. The leaves are usually shorter than the culms, a characteristic that is fairly useful in distinguishing this species from the similar but greenish spiked *C. vulpinoidea*.

9. **Carex bromoides** Schkuhr in Willd. Sp. Pl. 4:258. 1805. Fig. 11.

Plants perennial, densely cespitose, from long, thin, blackish, fibrillose rootstocks; culms sharply triangular, the sides concave, scabrous on the angles above, to 70 cm tall, equaling or much exceeding the leaves; leaves 3–5, 1.0–2.2 mm wide, stiff, the margins scabrous, the old leaves persistent; sheaths tight, with the hyaline ventral band cartilaginous and thickened at the concave summit; spikes 2–7 per culm, appressed-ascending, gynecandrous, all overlapping at the tip of the culm, forming a loose head 2–6 cm long; bracts scalelike with rough awns, only the lowest occasionally longer than the spike; pistillate scales lanceolate, acuminate to short-aristate, pale or soon suffused with amber, just reaching the base of the beak of the perigynium and as wide as the perigynium; perigynia 4–12 per spike, 3.5–4.5 mm long, 0.75–1.25 mm wide, appressed, ascending, lanceoloid, plano-convex, membranous except for the spongy base, substipitate, nerved dorsally and ventrally, greenish to light brown, the beak 1.25–1.50 mm long, bidentate, obliquely cut dorsally, serrulate; achenes lenticular, 1.50–1.75 mm long, 0.75 mm wide, substipitate, brown, jointed to a basally elongated style, positioned in the lengthwise center of the body of the perigynia; stigmas 2, elongated, flexuous to intertwined, reddish. April–May.

Low woods, seep springs, swamps, prairie bogs.

IL, IN, MO (FACW+), KY, OH (FACW).

Bromelike sedge.

This species is recognized by its dense clumps of very narrow leaves and its extremely narrow perigynia that are less than 1 mm wide. The somewhat similar but never aquatic *C. deweyana* lacks nerves on the convex face of the perigynia, which are also at least 1.5 mm wide.

10. **Carex brunnescens** (Pers.) Poir. in Lam. Encycl. Suppl. 3:286. 1813. Fig. 12.
Carex curta Good. var. *brunnescens* Pers. Syn. 2:539. 1807.
Carex canescens L. var. *sphaerostachya* Tuckerm. Enum. Meth. 10, 19. 1843.
Carex sphaerostachya (Tuckerm.) Dewey, Am. Journ. Sci. 49:44. 1845.
Carex canescens L. var. *vulgaris* Bailey, Bot. Gaz. 13:86. 1888.
Carex brunnescens (Pers.) Poir. var. *gracilior* Britt. in Britt. & Brown, Ill. Fl. 1:351. 1896.
Carex brunnescens (Pers.) Poir. var. *sphaerostachya* (Tuckerm.) Kukenth. Pflanzenr. 38, 4, Fam. 20:220. 1909.

Plants perennial, cespitose, from short, blackish, fibrillose rootstocks; culms triangular, scabrous, slender, weak, to 50 cm tall, longer than the leaves; leaves to 2 mm wide, lax, flat, dark green; sheaths tight, hyaline on the ventral band, with russet spots; spikes 3–6 per culm, gynecandrous, not overlapping, 4–5 mm long, in a more or less flexuous inflorescence up to 4 cm long; lowest bract seatceous; pistillate scales ovate, obtuse to acute, 3-nerved and green in the center, shorter than the

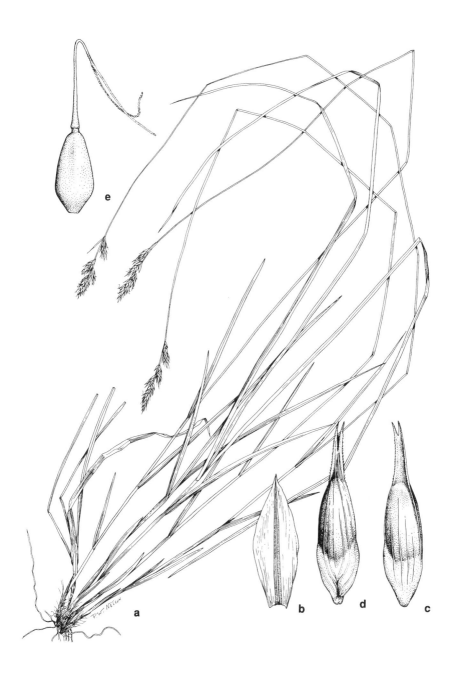

11. *Carex bromoides.*
a. Habit.

b. Pistillate scale.
c. Perigynium, dorsal view.

d. Perigynium, ventral view.
e. Achene.

12. _Carex brunnescens._
a. Habit.
b. Spike.

c. Pistillate scale.
d. Perigynium, dorsal view.
e. Perigynium, ventral view.

f. Achene.

perigynia; perigynia 5–10 per spike, broadly ellipsoid, 2.2–2.5 mm long, 1.0–1.5 mm broad, plump, tapering to a short, serrulate beak, spongy at base, nerveless or nearly so on the flat, ventral face, nerved on the convex dorsal face; achenes lenticular, ellipsoid, 1.3–1.5 mm long, 1.0 mm wide, yellow-brown, substipitate, apiculate, jointed to the style; stigmas 2, reddish brown. May.

Alkaline bogs.

IL, IN, OH (FACW).

Brown sedge.

This is a very rare species in the central Midwest. The plants from this area differ from typical var. *brunnescens* by its weaker leaves and generally smaller spikes. Those wishing to give a trinomial to our variety would call it var. *sphaerostachya*.

This is the only gynecandrous *Carex* in the central Midwest that has 3–6 non-overlapping spikes with perigynia only 1 mm wide.

11. **Carex buxbaumii** Wahlenb. Sv. Vet. Akad. Handl. 24:163. 1803. Fig. 13.

Plants perennial, cespitose, with long, horizontal rhizomes; culms up to 1 m tall, firm, slender, scabrous on the angles, at least beneath the inflorescence, red-purple at the base; leaves up to 4 mm wide, flat but with revolute margins, the lower surface glaucous and papillate, scabrous along the margins; lowest sheaths red-brown, fibrillose, upper sheaths with a hyaline ventral band russet-spotted, the ligule longer than wide; terminal spike gynecandrous, less commonly entirely staminate, up to 4 cm long, the lateral spikes 1–4, pistillate, erect to ascending, sessile or subsessile, the upper ones overlapping, up to 2 cm long; pistillate scales ovate, aristate, dark red-purple with a green midvein, a little longer than the perigynia; perigynia up to 40 per spike, ovoid, biconvex, 2.5–4.0 mm long, glaucous-green, papillate, 2-ribbed and several-nerved, substipitate, minutely beaked; achenes trigonous, 1.5–1.8 mm long, brown, puncticulate, apiculate, stipitate; stigmas 3.

Marshes, wet prairies, swales, usually in calcareous areas.

IL, IN, KS, KY, MO, OH (OBL).

Buxbaum's sedge.

This is the only species of *Carex* with dark red-purple, aristate pistillate scales and minutely beaked, glaucous perigynia.

12. **Carex canescens** L. Sp. Pl. 974. 1753. Fig. 14.
Carex canescens L. var. *subloliacea* Laestad. Nov. Act. Soc. Sci. Ups. 11:282. 1839.
Carex canescens L. var. *disjuncta* Fern. Proc. Am. Acad. 37:488. 1902.

Plants perennial, cespitose, from short, blackish, fibrillose rootstocks; culms sharply triangular, scabrous only beneath the head, to 80 cm tall, with last year's leaves persisting at base; leaves 5–8, 2–4 mm wide, lax, flat, glaucous to pale green, pale papillate beneath, the margins scabrous toward the tip, usually shorter than the culms; sheaths tight, clear on the hyaline ventral band, the lower sheaths brownish, the nerves darker, the ligule longer than broad; spikes 4–8 per culm, ovoid-cylindric, gynecandrous, overlapping or the lowest somewhat remote, 4–12 mm long, 3–5 mm wide, forming a loose or sometimes crowded head 2–10 cm long; bracts filiform to setaceous; pistillate scales broadly elliptic, obtuse to acuminate,

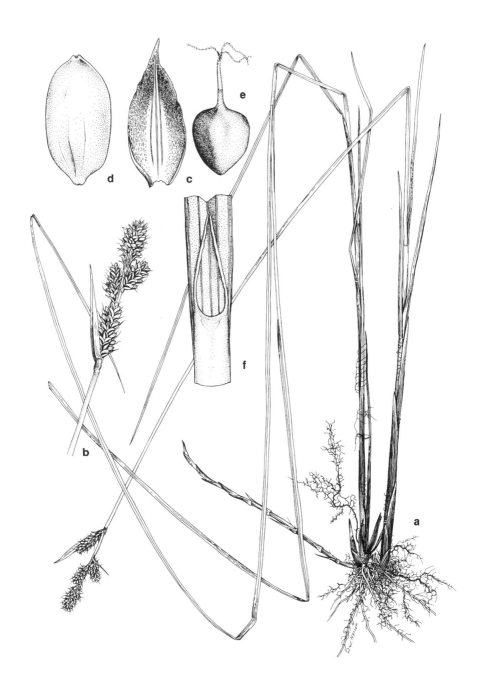

13. *Carex buxbaumii.*
a. Habit.
b. Inflorescence.

c. Pistillate scale.
d. Perigynium.
e. Achene.

f. Sheath with ligule.

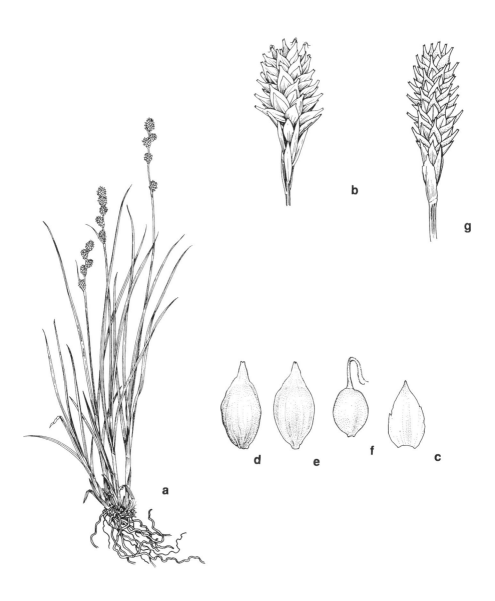

14. *Carex canescens.*
a. Habit.
b. Spike.

c. Pistillate scale.
d. Perigynium. dorsal view.
e. Perigynium. ventral view.

f. Achene.
g. Spike.

hyaline to brownish, with a 3-nerved green center, shorter than the perigynia; perigynia 10–30 per spike, 1.8–3.0 mm long, 1.25–1.75 mm wide, speading to ascending, narrowly ovoid to ellipsoid, plano-convex, membranous except for the spongy base, stipitate, obscurely nerved, pale green to pale brown, the beak 0.2–0.8 mm long, entire or emarginate, smooth or serrulate; achenes lenticular, about 1.5 mm long, 0.9 mm wide, ellipsoid, yellow-brown, substipitate, apiculate, jointed to the style; stigmas 2, reddish brown. April–May.

Sphagnum bogs.

IL, IN, OH (OBL).

Silvery sedge.

There are three varieties of this rare species of the central Midwest whose major range is to the north. Typical var. *canescens* has all of the spikes contiguous; var. *subloliacea* has some of its spikes remote and perigynia up to 2.2 mm long; var. *disjuncta* has some of its spikes remote and perigynia 2.2–3.0 mm long.

13. **Carex comosa** Boott, Trans. Linn. Soc. 20:117. 1846. Fig. 15.
Carex pseudocyperus L. var. *americana* Hochst. ex Bailey, Mem. Torrey Club 1:54. 1869.

Plants perennial, densely cespitose, from short, stout rhizomes; culms to 1.5 cm tall, stout, triangular, usually scabrous beneath the inflorescence, pale brown at base with several of last year's leaves persistent; leaves up to 17 mm wide, septate-nodulose, glabrous, pale green, scabrous on the margins; sheaths pale yellow with hyaline margins, concave at the mouth, the ligule longer than wide; lower bracts leaflike; terminal spike 1, staminate, rarely partly pistillate, up to 7 cm long, up to 7 mm thick, sessile or nearly so; staminate scales awned, reddish brown with a darker center; lateral spikes 3–6, pistillate, up to 5.5 cm long, up to 1.7 cm thick, at least the lower ones pendulous or on flexuous but stout peduncles; pistillate scales narrowly lanceolate, serrulate-awned, reddish brown with a darker center, about 1/2 as long as the perigynia; perigynia up to 100 or more per spike, crowded, divaricately spreading to reflexed, lanceoloid, 5–7 mm long, up to 1.5 mm wide, not inflated, glabrous, strongly nerved, yellow-green, coriaceous, tapering to a bidentate beak 1.5–2.0 mm long, the teeth stiff, recurved-spreading, 1.2–2.0 mm long; achenes trigonous, 1.7–2.0 mm long, brown, continuous with the persistent, flexuous style; stigmas 3, pale brown. June–August.

Swamps, sloughs, marshes, around ponds and lakes, wet meadows.

IA, IL, IN, MO, NE, OH (OBL).

Porcupine sedge.

Carex comosa differs from the similar *C. hystericina* in having some of its perigynia reflexed in the spike and in having teeth of its perigynia that are 1.2–2.0 mm long. It differs from *C. retrorsa* by its lower spikes that are pendulous rather than ascending.

14. **Carex conoidea** Schkuhr ex Willd. Sp. Pl. 4:280. 1805. Fig. 16.
Carex illinoensis Dewey, Am. Journ. Sci. II. 6:245. 1848.

Plants perennial, densely cespitose in tussocks, from short rhizomes; culms slender, triangular, to 50 cm tall, scabrous on the angles, brownish at the base; sterile shoots common; leaves up to 3 per culm, 3–5 mm wide, scabrous on the

15. *Carex comosa.*

a. Habit.
b. Pistillate scale.

c. Perigynium.
d. Achene.

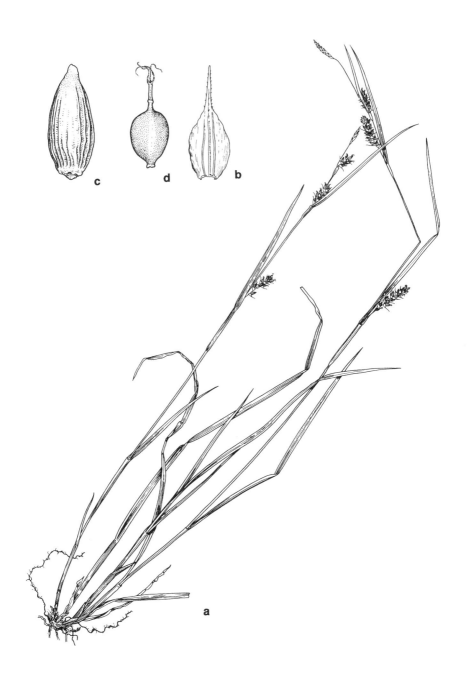

16. *Carex conoidea.*   a. Habit.
                             b. Pistillate scale.

c. Perigynium with scale.
d. Achene.

margins and the veins; sheaths red-dotted, at least where hyaline; terminal spike staminate, to 2 cm long, to 3 mm thick, on a scabrous peduncle up to 3 cm long, less commonly subsessile; staminate scales obtuse to acute, brownish with a green center; lateral spikes 1–3, pistillate, to 2.5 cm long, to 5 mm thick, sessile or on scabrous peduncles up to 2 cm long; bracts leaflike, often longer than the inflorescence; pistillate scales ovate, awned, brownish with a green center, shorter than to slightly longer than the perigynia; perigynia up to 25 per spike, oblongoid, 3–4 mm long, shiny, stramineous, finely impressed-nerved, glabrous, beakless or with a very short, entire beak; achenes trigonous, 2.0–2.2 mm long, yellow-brown, apiculate, stipitate; stigmas 3. April–June.

Wet meadows, wet prairies, uncommonly in shallow water.

IA, IL, IN (FACW+), OH (FACU).

Sedge.

The following group of characters distinguishes this species: long-stalked staminate spike; upper bracts usually longer than the inflorescence; perigynia with numerous, impressed nerves. The similar *C. crawei* has narrower leaves that are glaucous, pistillate scales that are red-brown and merely acute to cuspidate, and strongly nerved perigynia.

15. **Carex crawei** Dewey, Am. Journ. Sci. II. 2:246. 1846. Fig. 17.
Carex heterostachya Torr. Am. Journ. Sci. II. 2:248. 1846.

Plants perennial, with solitary culms arising from elongated rhizomes; culms up to 40 cm tall, obscurely triangular, smooth; leaves often in tufts, glaucous, sometimes folded, 1.5–3.5 mm wide, rather stiff, scabrous or smooth along the margins, shorter than the inflorescence; sheaths smooth, each cauline one with a pistillate spike; terminal spike staminate, up to 3 cm long, up to 3 mm thick, elevated on a scabrous peduncle up to 10 cm long; staminate scales obtuse, red-spotted; lateral spikes 2–4, pistillate, remote, up to 3 cm long, up to 6 mm thick, the uppermost often sessile, the others pedunculate, the lowest often near the base of the plant; pistillate scales acute to cuspidate, reddish brown with a green center, not reaching the beak of the perigynium; perigynia up to 45 per spike, ovoid, 3.0–3.5 mm long, 1.3–2.0 mm wide, glabrous, reddish punctate above the middle, strongly severalnerved, beakless or with a very short, hyaline beak; achenes trigonous, 1.7–2.0 mm long, yellow-brown, apiculate, substipitate; stigmas 3. April–June.

Fens, wet calcareous prairies, sandy flats, seldom in shallow water.

IA, IL, IN, KS, KY, NE, OH (FACW).

Crawe's sedge.

Although similar in appearance to *C. conoidea*, *C. crawei* may be recognized by its long-stalked staminate spike, its red-dotted perigynia, and its remote pistillate spikes, one of which is usually near the base of the plant.

16. **Carex crinita** Lam. Encycl. 3:393. 1791.

Plants perennial, cespitose, from stout rootstocks and slender stolons; culms to 1.5 m tall, stout, scabrous on the angles, often reddish near the base; leaves 3–5 per culm, up to 1.4 cm wide, somewhat scabrous along the slightly revolute margins,

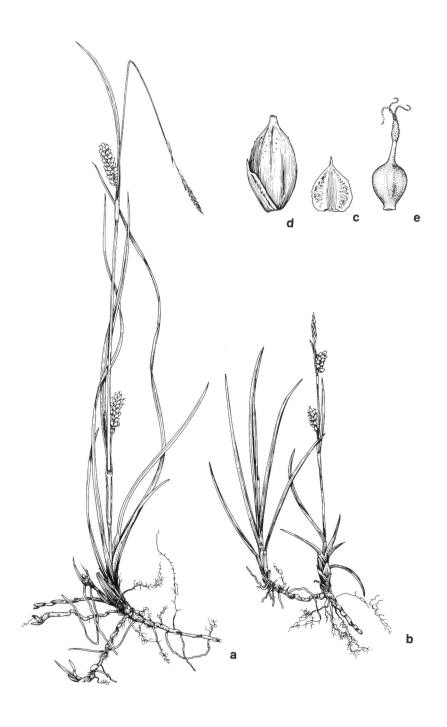

17. *Carex crawei.*

a, b. Habit.
c. Pistillate scale.

d. Perigyniuim with scale.
e. Achene.

septate between the nerves, the uppermost leaves longer than the culms; lowest sheaths red-brown to purple, smooth, strongly nerved and becoming fibrillose at maturity, the hyaline ventral bands thin; uppermost 1–3 spikes staminate, or with some pistillate flowers at the tip, up to 6 cm long, at maturity pendulous on slender peduncles; staminate scales acuminate to awned, conspicuous; lower 2–6 spikes pistillate, or with a few staminate flowers at the tip, narrowly cylindric, up to 12 cm long, pendulous on slender, smooth peduncles; lowest bracts foliaceous; pistillate scales ovate, retuse, abruptly rough-awned, red or yellow-brown with a green center, much longer than the perigynia; perigynia numerous, somewhat spreading, 2–4 mm long, ellipsoid to ovoid to suborbicular, green or pale brown or stramineous, lustrous, inflated, nerveless, beakless or with a beak 0.25 mm long, substipitate; achenes lenticular, 1.5–1.7 mm long, brownish, often slightly crimped on one side at the middle; stigmas 2. May–August.

Along streams, around ponds and lakes, marshes, fens, sloughs, swamps.

IA, IL, IN, MO (FACW+), KY, OH (OBL), NE (NI).

Fringed sedge.

This species is readily distinguished because all spikes are pendulous.

Two variations may be found in the central Midwest.

a. Lowest 2–6 spikes entirely pistillate; perigynia 2.0–3.5 mm long, often crimped on one side .... 16a. *C. crinita* var. *crinita*
a. Lowest 2–6 spikes pistillate but some of them with a few staminate flowers at apex; perigynia 3–4 mm long, not crimped .... 16b. *C. crinita* var. *brevicrinis*

16a. **Carex crinita** Lam. var. **crinita** Fig. 18 e.

16b. **Carex crinita** Lam. var. **brevicrinis** Fern. Rhodora 48:54. 1952. Fig. 18 a–d.

17. **Carex cristatella** Britt. in Britt. & Brown, Ill. Fl. 1:357. 1896. Fig. 19.
Carex cristata Schw. Ann. Lyc. N.Y. 1:66. 1824, misapplied.
Carex cristatella Britt. var. *catelliformis* Farw. Papers Mich. Acad. 2:17. 1923.
Carex cristatella Britt. f. *catelliformis* (Farw.) Fern. Rhodora 44:284. 1942.

Plants perennial, densely cespitose, from short, brownish black, fibrillose rootstocks; culms to 90 cm tall, sharply triangular with concave sides (pressing nearly flat), the angles scabrous beneath the inflorescence, stiff, usually shorter than the culms, with sterile culms common; leaves 3–6 per fertile culm, ascending to spreading, 3.0–7.5 mm wide, often revolute, green, the margins scabrous; sheaths relatively loose, wing-margined, or inflated near summit; spikes 4–12 per culm, less than 1 cm in diameter, green, globose, gynecandrous, abruptly tapering to the base, usually crowded in a headlike inflorescence up to 4 cm long, rarely separated into a moniliform inflorescence; lowest bracts setaceous, scabrous, upper bracts scalelike; pistillate scales lanceolate, acute to acuminate, three-fourths as long as and narrower than the perigynia, often hidden by the recurving perigynia, the margins tan- to brown-hyaline, the center darker and 3-nerved; perigynia many per spike, 3.5–4.0 mm long, 1.4–1.8 mm wide, lance-ovate, widest at the middle, but with the body often

18. *Carex crinita*.
a. Habit.

b. Pistillate scale.
c. Perigynium.

d. Achene.
e. Pistillate scale.

a–d are var. *brevicrinis*. e is var. *crinita*.

19. *Carex cristatella.*
a. Habit.

b. Pistillate scale.
c. Perigynium, dorsal view.

d. Perigynium, ventral view.
e. Achene.

suborbicular, the tips ascending to widely spreading, plano-convex, sometimes distended over the achenes dorsally and ventrally, membranous, distinctly nerved dorsally and ventrally, the wing abruptly narrowed just above the base, the beak 1.0–1.5 mm long, often twisted and constricted at the base, serrulate, bidentate, the teeth appressed; achenes lenticular, 1.3–1.6 mm long, 0.5–0.8 mm wide, apiculate, stipitate, weakly jointed to the deciduous style; stigmas 2, slender, elongated, reddish brown. May–July.

Marshes, wet woods, swales, streambanks, ditches, wet meadows, bogs.

IA, IL, IN, MO (FACW+), KS, KY, NE, OH (FACW).

Round-spikelet sedge.

This species is distinguished by its globose spikes, outward curving perigynia that obscure the pistillate scales, and perigynia less than 2 mm wide that taper to the base. Specimens in which the spikes are separated to form a moniliform inflorescence may be designated f. *catelliformis.* These plants resemble *C. projecta*, but the lowest spikes are not clearly separated as they are in *C. projecta.*

18. **Carex crus-corvi** Shuttlw. in Kunze, Riedgr. Suppl. 128. 1842. Fig. 20.
Carex siccaeformis Boott, Journ. Bost. Nat. Hist. Soc. 5:113. 1845.
Carex halei Dewey, Am. Journ. Sci. 2:248. 1846.

Plants perennial, cespitose, with fibrous roots and short, stout, blackish brown rhizomes; culms erect, stout, up to 1.2 m tall, strongly 3-angled and winged, the sides concave, not pressing completely flat, the angles rough to the touch, light brown at base, usually overtopped by the leaves; leaves 4–8 per culm, gray-green, up to 90 cm long, 5–12 mm wide, flat, pale green, septate-nodulose, rough along the margins; sheaths open, loose, concave at the summit, purple-dotted, septate-nodulose, the ventral side smooth, broadly white- to tan-hyaline, tearing easily; ligule wider than long; inflorescence compound, usually conspicuously branched, up to 20 cm long, up to 6 cm broad; bracts setaceous to scalelike to lacking, up to 8 cm long, rough-awned, rarely as long as the lowest branch of spikes; spikes 15–25, the staminate flowers above the pistillate; pistillate scales lanceolate to ovate, acuminate to aristate, green to pale brown to hyaline on the margins, the center green and 3-nerved, shorter than the perigynia; staminate scales narrowly lanceolate, pale brown; perigynia lanceoloid to subuloid, 6–8 mm long, 1.5–2.2 mm wide, plano-convex, swollen and spongy thickened at base, stipitate, yellow to brown, conspicuously nerved on the dorsal face, nearly nerveless ventrally, tapering to a long, subulate, serrulate, 2-toothed beak 4–6 mm long and about twice as long as the body of the perigynia; achenes 1.8–2.5 mm long, 1.25–1.50 mm wide, lenticular, ovoid, long-apiculate, substipitate, jointed to the style; stigmas 2, yellowish brown. May–July.

Swamps, sloughs, ditches, along rivers and streams.

IA, IL, IN, KY, MO, NE, OH (OBL).

Cock-spur sedge.

The perigynia are some of the most distinctive in the genus. They are greatly enlarged and spongy at the base, tapering to long-subulate beaks much longer than the body of the perigynia. The leaf sheaths are septate-nodulose and purple-spotted.

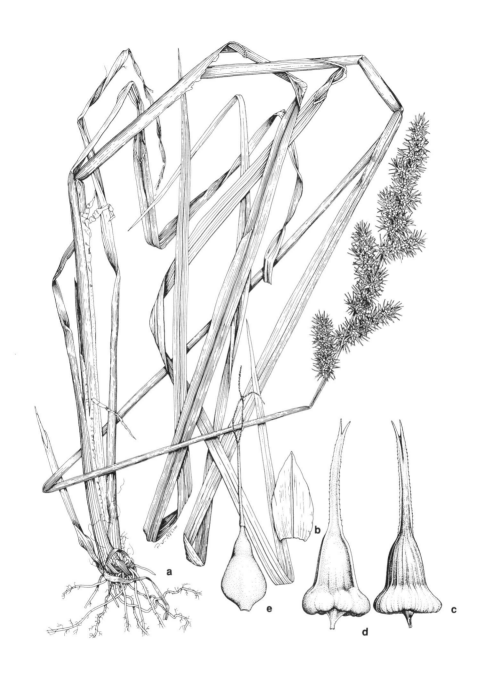

20. *Carex crus-corvi.*
a. Habit.

b. Pistillate scale.
c. Perigynium, dorsal view.

d. Perigynium, ventral view.
e. Achene.

19. **Carex cryptolepis** Mack. Torreya 14:156. 1914. Fig. 21.
Carex flava L. var. *fertilis* Peck, N.Y. State Mus. Rep. 48:197. 1896.

Plants perennial, cespitose, with short rhizomes; culms to 50 cm tall, 3-angled, usually smooth, pale brown at the base; sterile shoots usually conspicuous; leaves up to 4.5 mm wide, light green, slightly scabrous along the margins; sheaths convex at the summit; terminal spike staminate, or occasionally with a few perigynia at the tip, sessile or on short peduncles; staminate scales acute, yellowish with a green center; pistillate spikes 2–5, globose or ovoid, crowded near tip of culm, with one other often remote toward middle of culm, 6–10 mm thick, each subtended by a leafy bract; pistillate scales ovate, acute to acuminate, pale green to stramineous, the tip not reaching the base of the beak of the perigynium; perigynia up to 35 per spike, 3.2–4.5 mm long, narrowly ovoid, greenish, spreading, or the lower ones reflexed, shiny, few-nerved, with a beak nearly as long as the body, the beak bidentate, greenish, smooth; achenes trigonous, with concave sides, nearly black, shiny, 1.3–1.5 mm long, short-apiculate; stigmas 3. May.

Fens.

IL, IN, OH (OBL).

Yellow sedge.

This species and *C. flava* are the only species of *Carex* with globose, yellowish, pistillate spikes in which several of the lower perigynia are reflexed. The very similar *C. flava* has wider, yellow-green leaves, longer perigynia with serrulate beaks, and conspicuous pistillate scales.

20. **Carex decomposita** Muhl. Descr. Gram. 264. 1817. Fig. 22.

Plants cespitose; rhizomes woody, thick, blackish, fibrous; roots wiry; culms mostly fertile, spreading and arching, smooth or roughened, terete or obtusely angled, 4–8 mm wide near base, up to 1.5 m tall, slightly shorter to slightly longer than the leaves, usually with several old leaves persistent near the base, producing leaves in the lower third; leaves up to 75 cm long, up to 8 mm wide, flat or canaliculate, stiff, the margins and main nerves serrulate; sheaths tight, septate-nodulose, pale, sometimes red-dotted, concave at the usually reddish mouth; inflorescence up to 18 cm long, compound, the lower branches up to 4 cm long; bracts absent or setaceous, up to 5 cm long; spikes crowded, the staminate flowers above the pistillate; pistillate scales triangular-ovate, mucronate, up to 2.8 mm long, about as long as the perigynia, hyaline, with a 3-nerved green center, persistent after the perigynia have fallen; staminate scales apical, inconspicuous, crowded; perigynia obovoid, crowded, coriaceous, biconvex, shiny, spreading, 2.0–2.8 mm long, 1.50–1.75 mm broad, spongy at base, olive to blackish, obscurely few-nerved above, strongly several-nerved below, rounded at summit except for the beak, tapering to a stipitate base, the beak 0.5 mm long, flat, serrulate, bidentate; achenes about 1 mm long, closely enveloped by the perigynia, lenticular, apiculate, stipitate, jointed with the short style; stigmas 2, short, light reddish brown. April–July.

Swamps, sinkhole ponds.

IL, IN, KY, MO, OH (OBL).

Cypress-knee sedge.

21. *Carex cryptolepis.*

a. Habit.
b. Pistillate scale.

c. Perigynium.
d. Achene.

22. *Carex decomposita.*
a. Habit.

b. Pistillate scale.
c. Perigynium, dorsal view.

d. Perigynium, ventral view.
e. Achene.

This species grows primarily on the swollen bases of trees and on fallen logs in swamps, often those with bald cypress trees present. This is the only species of *Carex* with a compound inflorescence and black or olive perigynia.

21. **Carex diandra** Schrank, Cent. Bot. Ammerk. 57. 1781. Fig. 23.
Carex teretiuscula Gooden, Trans. Linn. Soc. 2:163. 1794.
Carex teretiuscula Gooden var. *major* Koch, Syn. Fl. Germ. 751. 1837.

Plants loosely cespitose; rhizomes short, slender, blackish, fibrous-scaly; roots wiry, abundant; culms mostly fertile, erect, becoming slightly spreading, roughened on the rounded angles, up to 1.1 m tall, exceeding the leaves, brown or dark brown at base, producing leaves in the lower fourth; blades 4–5 per culm, up to 30 cm long, 1–3 mm wide, flat, canaliculate or plicate, ascending or spreading, the margins and sometimes the tip scabrous, light green; sheaths tight, membranous and hyaline ventrally, striate dorsally, strongly red-dotted, truncate or convex at the mouth; inflorescence up to 8 cm long but usually shorter, up to 12 mm thick, subcylindric, continuous, compound and paniculate at the lowest node; bracts subulate, up to 1 mm long or absent; spikes several, crowded, dark brown, the staminate flowers above the pistillate; pistillate scales ovate-lanceolate, up to 3.5 mm long, more or less concealing the sides of the perigynia, acute to short-cuspidate, greenish but tinged with brown, the margins scarious above the middle; staminate scales apical, inconspicuous, very narrow; perigynia ovoid, biconvex, strongly rounded on the back, ascending to widely spreading, 2–3 mm long, about 1 mm broad, strongly nerved on the lower side, dark brown to olive-black, shiny, stipitate, tapering to a green beak, the beak up to 1 mm long, apiculate, stipitate, jointed to a short style; stigmas 2, short, reddish brown. May–July.

Swamps, wet meadows, shores, occasionally in standing water.

IA, IL, IN, NE, OH (OBL).

Two-anthered sedge.

Carex diandra and *C. prairea* are very similar in appearance, but *C. diandra* has more spreading, shiny, blackish perigynia and leaves that are never revolute.

22. **Carex disperma** Dewey, Am. Journ. Sci. 8:266. 1824. Fig. 24.
Carex tenella Schk. Riedgr. 23, pl. 104. 1801. non Thuill. (1844).

Plants perennial, loosely cespitose to mat-forming, with fibrous roots and usually slender, brown, scaly rhizomes; culms weakly erect to reclining, to 60 cm long, very slender, about 0.5 mm wide, sharply triangular, rough to the touch beneath the inflorescence, light brown at base, with several old leaves persistent at base; sterile culms common; leaves 3–6, up to 30 cm long, 1–2 mm broad, soft, dark green, rough to the touch along the margins and veins beneath, dark green, usually shorter than the culms; sheaths open, tight, concave at the summit, hyaline; ligule wider than long; inflorescence composed of a few interrupted spikes, up to 25 mm long, up to 5 mm broad; bracts setaceous to scalelike, rough-awned, enlarged at base, up to 1 cm long; spikes 2–7 per culm, aggregated above, the lower 2–4 remote, the 1–3 staminate flowers above the pistillate flowers; pistillate scales ovate, acute to acuminate to aristate, hyaline with a green midvein, shorter than to equal the

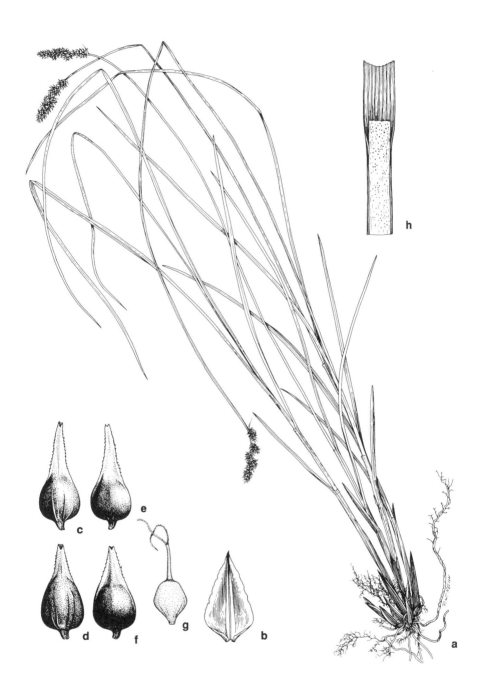

23. *Carex diandra*.
a. Habit.
b. Inflorescence.

c, d. Perigynia, dorsal views.
e, f. Perigynia, ventral views.
g. Achene.

h. Sheath.

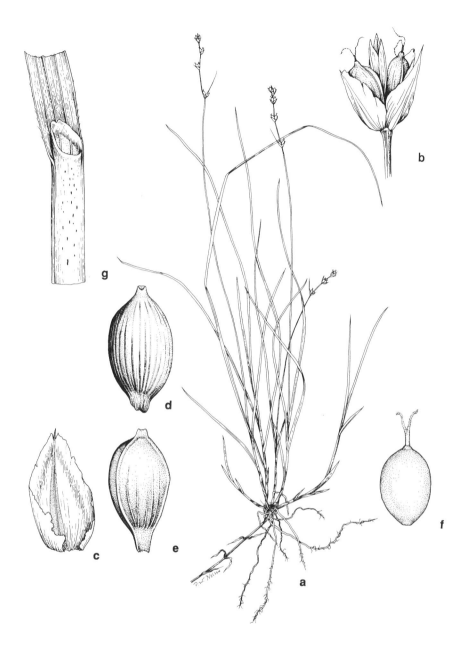

24. *Carex disperma.*
a. Habit.
b. Spike.

c. Pistillate scale.
d. Perigynium, dorsal view.
e. Perigynium, ventral view.

f. Achene.
g. Sheath with ligule.

length of the perigynia, narrower than the perigynia; staminate scales narrowly lanceolate; perigynia 1–6 per spike, ellipsoid-ovoid, 1.75–3.00 mm long, 1.0–1.5 mm broad, unequally biconvex, olive to yellow-green, white-puncticulate, slightly spongy at base, substipitate, the margins rounded and unwinged, finely nerved, rounded at the apex and with 2 entire teeth less than 0.3 mm long; achenes 1.5–2.0 mm long, 1 mm wide, lenticular, yellow-brown, shiny, jointed to the style; stigmas 2, reddish brown, slender. May–August.

Marshes, swamps.

IL, IN (OBL), OH (FACW+).

Soft-leaved sedge.

This weak species is sometimes mat-forming. The sparsely flowered spikes and the ellipsoid-ovoid, biconvex, nearly beakless perigynia are also diagnostic. *Carex trisperma* is similar, but it has staminate flowers below the pistillate ones.

23. **Carex echinata** Murray, Prod. Stirp. Gott. 76. 1770. Fig. 25.
Carex leersii Willd. Fl. Berol. Prod. 29, 1787, *nomen illeg.*
Carex stellulata Goodenough, Trans. Linn. Soc. 2:144. 1794.
Carex echinata Murray var. *cephalantha* Bailey, Mem. Borry Club 1:58. 1889.
Carex stellulata Goodenough var. *cephalantha* (Bailey) Fern. Rhodora 4:222. 1902.
Carex cephalantha (Bailey) Bicknell, Bull. Torrey Club 35:493. 1908.

Plants perennial, cespitose, from short rhizomes; culms 3-angled, slightly sca-brous, to 90 cm tall; leaves 2–6, 1.0–3.5 mm wide, plicate, green, scabrous along the margins, a little shorter than the culms; sheaths tight, smooth, the inner band hyaline and sometimes purple-dotted, concave at the apex, the lower ones light brown; inflorescence up to 7.5 cm long, with 3–8 crowded or separate, sessile spikes; terminal spike 0.5–2.0 cm long, gynecandrous, the staminate part up to 1.5 cm long, 2- to 17-flowered, the pistillate part up to 1 cm long, 4- to 26-flowered; lateral spikes up to 1.5 cm long, gynecandrous, with scalelike bracts; pistillate scales ovate, 1.5–3.0 mm long, castaneous, with a green center and hyaline margins, acute at the tip, reaching the base of the beak of the perigynium; perigynia 2.5–4.5 mm long, 1–2 mm wide, lanceoloid to ovoid, plano-convex, the ventral surface with up to 12 veins, or nerveless, spongy-thickened at the base, green to dark brown, sessile, tapering to a beak, with the lower perigynia spreading to reflexed; beak 1–2 mm long, serrulate, with sharp, stiff teeth at the apex; achenes biconvex, 1.5–2.0 mm long, 1.0–1.5 mm broad, substipitate, jointed to the deciduous style; stigmas 2. May–June.

Wet meadows, usually not in standing water.

IA, IL, IN, OH (OBL).

Star sedge.

In *Carex echinata*, *C. atlantica*, and *C. interior*, the perigynia are arranged in a spike to resemble a star. Some of the perigynia in *C. echinata* are usually more than 3 mm long, while the perigynia in *C. atlantica* and *C. interior* are usually less than 3 mm long, although in *C. atlantica* a few specimens have perigynia up to 3.5 mm long.

Carex cephalantha is sometimes considered to be a separate species, but I am con-sidering it to be conspecific with *C. echinata*.

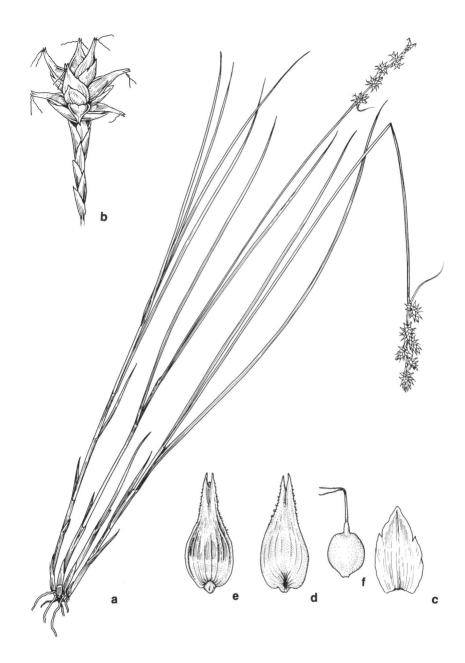

25. *Carex echinata.*
a. Habit.
b. Inflorescence.

c. Pistillate scale.
d. Perigynium, dorsal view.
e. Perigynium, ventral view.

f. Achene.

24. Carex emoryi Dewey in Torr. Mex. Bound. Survey 230. 1859. Fig. 26.
Carex stricta Lam. var. *emoryi* (Dewey) Bailey, Proc. Am. Acad. Sci. 22:85. 1886.

Plants perennial, cespitose, with rather stout, brown, scaly rhizomes; sterile shoots common; culms to 1.1 m tall, stout, usually scabrous on the angles, red-purple at the base, often with last year's leaves persistent; leaves up to 6 mm wide, flat to revolute, scabrous along the margins, septate between the veins, usually longer than the culms; lowest sheaths bladeless, red to purple-brown to blackish, usually disintegrating into fibrillose filaments, convex at the mouth, the ligule shorter than the width of the leaf; upper 1–3 spikes staminate, up to 5 cm long, more or less erect, pedunculate; staminate scales obovate, rounded at the tip, brown with a tawny midvein; lowest 3–5 spikes pistillate, often with a few staminate flowers at the tip, up to 10 cm long, erect, subsessile to short-pedunculate; pistillate scales 2.0–3.5 mm long, ovate to elliptic, subacute to acuminate, hyaline to reddish brown with a tawny midvein, about as long as to surpassing the perigynia; perigynia numerous, flat or biconvex, elliptic to ovate, 1.7–3.2 mm long, rounded at the tip, dull green to stramineous, 3- to 5-nerved, usually granular-papillate with a minute beak up to 0.3 mm long; achenes lenticular, 1–2 mm long, dull brown; stigmas 2. April–July.

Along streams, ditches, wet meadows, often in standing water.

IA, IL, IN, KS, MO, NE, OH (OBL).

Emory's sedge.

Carex emoryi and *C. stricta* are very similar in appearance and in habitats. *Carex emoryi* differs in its very short ligules and fewer granular papillae on its perigynia. *Carex emoryi* also resembles *C. aquatilis* var. *substricta* but *C. emoryi* lacks leafy lower sheaths.

25. Carex festucacea Schk. in Willd. Sp. Pl. 4:242. 1805. Fig. 27.

Plants perennial, densely cespitose, from short, black, fibrillose rhizomes; culms to 1 m tall, stout, sharply triangular, scabrous on the angles, at least beneath the inflorescence, longer than the leaves, brownish black at the base with the old leaves often remaining as stubble; leaves 3–5 per culm, ascending, 1–5 mm wide, flat, green, the margins scabrous toward the apex; sheaths tight, green-nerved nearly throughout or with a very narrow hyaline ventral band, often septate, the mouth prolonged, the lowermost sometimes becoming purplish; spikes 3–10 per culm, 6–16 mm long, gray-green, obtuse, gynecandrous, strongly clavate at the base, the staminate flowers conspicuous, the spikes sometimes rather crowded or more often separated, sometimes in a moniliform inflorescence 2.5–6.0 cm long, the axis of the inflorescence not flexed but straight; lowest bract sometimes setaceous, scalelike, the upper bracts scalelike; pistillate scales lanceolate, acute to acuminate, reaching or surpassing the base of the beak of the perigynium, the center greenish and 3-nerved, the margins hyaline; perigynia up to 20 per spike, 2.5–3.5 mm long, 1.5–2.2 mm wide, oval to orbicular, widest nearer the base, spreading or ascending, plano-convex, green to stramineous, subcoriaceous, strongly nerved on the outer face, with 2–4 fine nerves on the inner face, winged to the base, substipitate, the beak 1.0–1.3 mm long, serrulate, bidentate; achenes lenticular, 1.0–1.6 mm long, 1.0–1.3 mm wide,

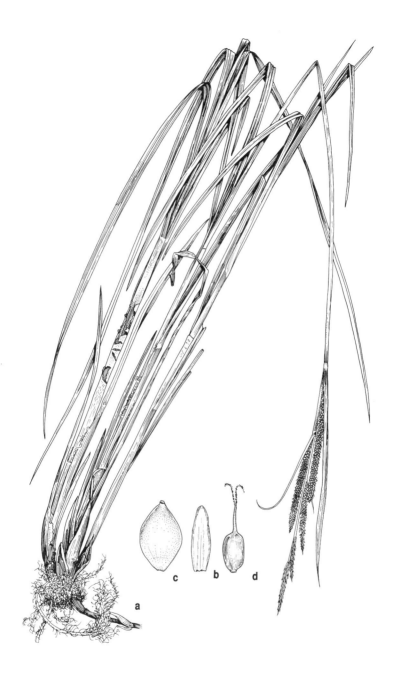

26. *Carex emoryi.*

a. Habit.
b. Pistillate scale.
c. Perigynium.
d. Achene.

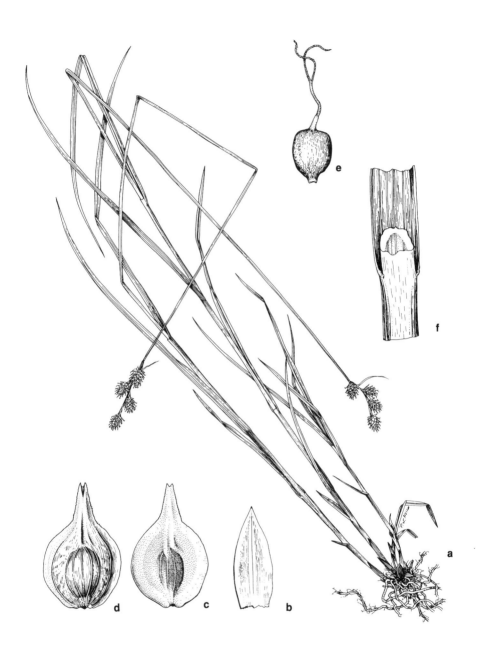

27. *Carex festucacea.*
a. Habit.
b. Pistillate scale.

c. Perigynium, dorsal view.
d. Perigynium, ventral view.
e. Achene.

f. Sheath with ligule.

apiculate, stipitate, weakly jointed with the deciduous style; stigmas 2, short, slender, brownish. April–June.

Moist prairies, moist savannas, low woods, occasionally in shallow water. IA, IL, IN, MO (FAC), KY, OH (FAC-), KS (FACW-).

Fescue sedge.

Several species are similar in appearance to *C. festucacea*, including *C. tenera, C. longii*, and *C. albolutescens*. Specimens with separated spikes, therefore resembling *C. tenera*, may be recognized by the strongly clavate base of the spikes and by the broadly ovate to suborbicular rather than narrowly ovate perigynia. Specimens with more crowded spikes tend to resemble *C. longii* and *C. albolutescens* but differ in having the perigynia broadest nearest the base rather than nearest the middle.

26. **Carex flava** L. Sp. Pl. 2:975. 1753. Fig. 28.

Plants perennial, cespitose, with short rhizomes; culms to 80 cm tall, 3-angled, usually smooth, pale brown at the base; sterile shoots usually conspicuous; leaves up to 6 mm wide, yellow-green, slightly scabrous along the margins; sheaths convex at the summit; terminal spike staminate, or occasionally with a few perigynia at the tip, sessile or on short peduncles; staminate scales acute, yellowish with a green center; pistillate spikes 2–6, ellipsoid to subglobose, crowded near tip of culm, with one other often remote toward middle of culm, 10–14 mm thick, each subtended by a leafy bract; pistillate scales lanceolate to ovate, acuminate, brown, the tip not reaching the base of the beak of the perigynium; perigynia up to 35 per spike, 4–6 mm long, narrowly ovoid to subuloid, yellowish, spreading, or the lower ones reflexed, shiny, few-nerved, with a beak 1.5–2.7 mm long, 1/2 to nearly as long as the body, the beak bidentate, greenish, smooth or serrulate; achenes trigonous, broadly obovoid, with concave sides, nearly black, shiny, 1.4–1.8 mm long, short-apiculate; stigmas 3. May.

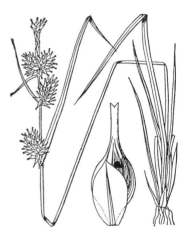

28. *Carex flava*. Habit and achene.

Bogs, fens.

KY, OH (OBL).

Yellow sedge.

Carex flava differs from the very similar *C. cryptolepis* by its yellow rather than greenish perigynia, its yellow-green rather than green leaves, and its larger pistillate spikes 10–14 mm across.

27. **Carex frankii** Kunth. Enum. Pl. 2:498. 1837. Fig. 29.
Carex stenolepis Torr. Ann. Lyc. N.Y. 3:420. 1836, non Wahlenb. (1803).

Plants perennial, cespitose, with short, stout rhizomes; culms to 80 cm tall, triangular, usually slightly rough to the touch, usually purplish at the base; sterile shoots usually present; leaves up to 10 mm wide, deep green, firm, slightly scabrous along the margins and usually on the veins; sheaths tight, septate-nodulose, yellow-brown, hyaline at the summit, the lowermost often red-tinged, the ligule about as wide as long; terminal spike staminate or less commonly with pistillate flowers at tip, up to 3 cm long, up to 3 mm thick; staminate scales hyaline with a green center, awned; pistillate spikes 3–7, cylindric, up to 4 cm long, up to 1.2 cm thick, sessile or on short peduncles; bracts many times longer than the inflorescence; pistillate scales setaceous, slightly scabrous, green, much longer than the perigynia; perigynia up to 100 per spike, obconic, truncate across the top, 4–5 mm long, up to 2.5 mm broad, olive-green, becoming brownish, many-nerved, abruptly contracted into a bidentate beak to 2.5 mm long; achenes trigonous, yellow-brown, about 1.5 mm long, substipitate; stigmas 3. May–September.

Wet ditches, along streams, moist woods, wet prairies, sometimes in shallow water. IA, IL, IN, KS, KY, MO, NE, OH (OBL).

Frank's sedge.

This sedge, which blooms for a longer period than most species of *Carex*, is readily distinguished by its extremely long bracts that subtend the inflorescence and by the achenes that are abruptly cut straight across at the top. On some specimens, the terminal spikes may contain nearly all pistillate flowers.

28. **Carex gigantea** Rudge, Trans. Linn. Soc. 7:99. 1804. Fig. 30.
Carex lacustris Willd. var. *gigantea* (Rudge) Pursh, Fl. Am. Sept. 1:45. 1814.
Carex grandis L. H. Bailey, Mem. Torrey Club 1:13. 1889.
Carex lupulina Muhl. var. *gigantea* (Rudge) Britt. Mem. Torrey Club 5:84. 1894.
Carex gigantea Rudge var. *grandis* (L. H. Bailey) Farwell, Rhodora 23:87. 1921.

Plants perennial, cespitose or growing singly, from fibrous roots, with long, connected rhizomes; culms up to 1.06 m tall, stout, much exceeded by the upper leaves and bracts, triangular, smooth throughout, light brown or green, becoming purplish near the base; leaves up to 60 cm long, 5–14 mm wide, tapering gradually to an attenuated tip, septate-nodulose, flat, dull green, with margins rough; sheaths little or not at all prolonged at the mouth, yellowish tinged to white-hyaline ventrally; staminate spikes 1–7, linear, 1.1–9.0 cm long, 1.5–4.0 mm broad, borne on short or long peduncles, with scales linear to lanceolate to obovate, strongly awned to acuminate, stramineous with a several-nerved green center and hyaline margins; bracts 1/2 as long to exceeding the spikes; pistillate spikes 2–6, oblongoid-cylindric, 1–8 cm long, 1.7–3.0 cm broad, erect, densely flowered, sessile or on peduncles up to 15 cm long; bracts leaflike, exceeding the inflorescence, with sheaths short-prolonged at the mouth; perigynia 9–92, diverging from the axis of the pistillate spikes at angles of 60–90 degrees; pistillate scales lanceolate, acuminate, or the lower awned, stramineous with a several-nerved green center and hyaline margins, much narrower than and usually shorter than the perigynia; perigynia subuloid, 8–16 mm long, 2–5 mm broad, round-truncate at the base, barely inflated, glabrous, deep green or brownish yellow at maturity, with 16–20 nerves, abruptly contracted or tapering into a well-defined beak usually more than 1/2 the length of the total perigynia, the beak bidentate, with teeth up to 1.5 mm long, smooth

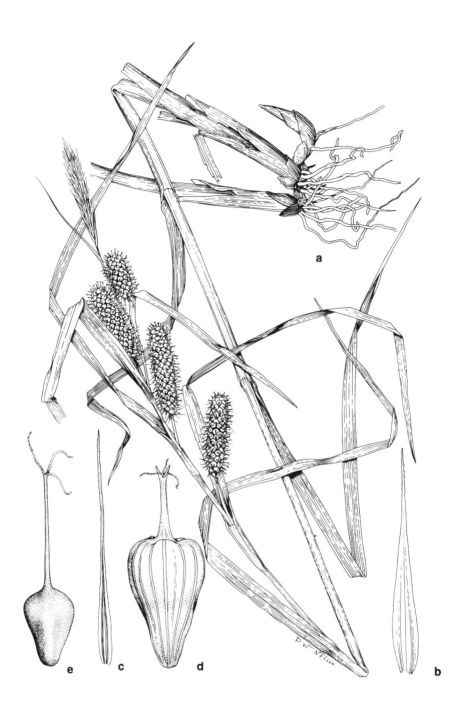

29. *Carex frankii.*
a. Habit.

b. Staminate scale.
c. Pistillate scale.

d. Perigynium.
e. Achene.

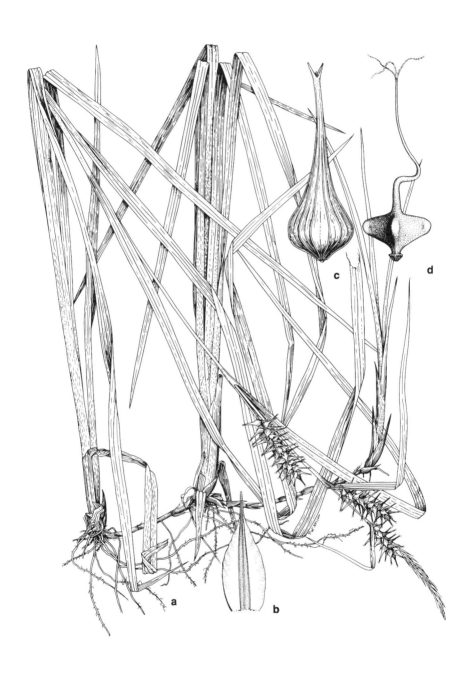

30. *Carex gigantea.*

a. Habit.
b. Pistillate scale.

c. Perigynium.
d. Achene.

within; achenes depressed, obovoid to ovoid, wider than long to as wide as long, 1.5–2.5 mm long, 2.0–3.5 mm broad, the sides deeply concave, the angles obtuse and flattened into broad winglike projections, abruptly contracted into a stipitate base, abruptly contracted into and continuous with the persistent, slender, barely flexuous to completely coiled style; stigmas 3, short, slender, blackish. May–September.

Swamps, bottomland forests, occasionally in shallow water.

IL, IN, KY, MO (OBL).

Knobby-fruited sedge.

The distinctive features of this species are the achenes that have the angles flattened into broad wings and the sides deeply concave.

29. **Carex glaucescens** Ell. Sketch. Bot. S. Carol. 2:553. 1824. Fig. 31.

Cespitose perennial from short, black rhizomes; culms to 1.2 m tall, glabrous; leaves 4–8 mm wide, glabrous; staminate spike 1, terminal, erect, to 4 cm long, 5–7 mm thick, on a short peduncle; pistillate spikes 3–6, some of them sometimes with a few staminate flowers at the apex, the uppermost spikes ascending and on short peduncles, the lowermost spikes nodding and on longer peduncles, up to 5 cm long, 7–9 mm thick, the lowest bract up to 4.5 mm wide, sheathless; pistillate scales about as long as the perigynia, obovate, red-brown with a hyaline margin and a green midnerve slightly exserted at the tip; perigynia 60–100 per spike, ellipsoid to obovoid, somewhat inflated, 3–4 mm long, with 2 strong nerves and 2–3 obscure nerves, purple-brown but glaucous, rounded at

31. *Carex glaucescens.* Habit, perigynium, and scale.

the base, tapering at the apex to an entire beak 0.2–0.3 mm long; achenes trigonous, apiculate, 2.5–3.5 mm long. April–June.

Marshes, wet woods, wet ditches.

KY (OBL).

Glaucous sedge.

This species resembles *C. joori*, but differs by its smaller perigynia that have two strong and two or three obscure nerves.

30. **Carex gynandra** Schwein. Ann. Lyc. N.Y. 1 (1):70. 1824. Fig. 32.
Carex crinita Lam. var. *gynandra* (Schwein.) Schwein. & Torr. Ann. Lyc. N.Y. 1:360. 1825.

Plants perennial, cespitose, from stout rootstocks and slender stolons; culms to 1.6 m tall, stout, scabrous on the angles, often reddish near the base; leaves 3–5 per culm, up to 1.2 cm wide, somewhat scabrous along the slightly revolute margins, septate between the nerves, the uppermost leaves longer than the culms; lowest sheaths red-brown to purple, strongly nerved and becoming fibrillose at maturity, scabrous, hispidulous, the hyaline ventral bands thin; uppermost 1–3 spikes staminate with some pistillate flowers at the tip, up to 6 cm long, at maturity pendulous on slender peduncles; staminate scales acuminate to awned, conspicuous; lower 2–6 spikes pistillate, or with a few staminate flowers at the tip, narrowly cylindric, up to 10 cm long, pendulous on slender, smooth peduncles; lowest bracts foliaceous; pistillate scales ovate, retuse, abruptly rough-awned, red or yellow-brown with a green center, much

32. *Carex gynandra.* Habit, perigynium, and scale.

longer than the perigynia; perigynia numerous, ascending, 3–4 mm long, ellipsoid to ovoid to suborbicular, green or pale brown or stramineous, lustrous, somewhat inflated, nerveless, beakless or with a beak 0.25 mm long, substipitate; achenes lenticular, 1.5–1.7 mm long, brownish, often slightly crimped on one side at the middle; stigmas 2. May–August.

Swampy woods, marshes, wet ditches.

OH (OBL), as *C. crinita* by the U.S. Fish and Wildlife Service.

Rough-sheathed fringed sedge.

There are two major characters that separate this species from the very similar *C. crinita.* All of the upper spikes have some pistillate flowers at their tip, and the lower sheaths are scabrous and hispidulous.

31. **Carex haydenii** Dewey, Am. Journ. Sci. II, 18:103. 1854. Fig. 33.
Carex stricta Lam. var. *haydenii* (Dewey) Kukenth. Pflanzenr. IV, 20:330. 1909.

Plants perennial, cespitose, with short, erect rhizomes; sterile shoots common; culms to 1.2 m tall, stout, scabrous on the angles, red-brown at the base, with last year's leaves persistent; leaves 2.5–5.0 mm wide, keeled, usually revolute, firm, scabrous on the margins, shorter than the culms; lowest sheaths bladeless, brown, sometimes fibrillose, the uppermost smooth with a hyaline ventral band, the ligule longer than the width of the leaf; upper 1–2 spikes staminate, erect, up to 4 cm long, subsessile to pedunculate; staminate scales 3–4 mm long, obovate, subacute, brown to red-brown; lower 2–3 spikes pistillate, erect, sometimes with a few staminate flowers at the tip, up to 5 cm long, sessile or short-pedunculate; pistillate scales 2.0–3.5 mm long, oblong to ovate, acute to acuminate, brown or red-brown with a green or tawny midvein, longer than the perigynia; perigynia numerous, biconvex but plump, 1.5–2.8 mm long, ellipsoid to obovoid, olive green to brown, papillate, red-dotted, inflated, nerveless, substipitate, minutely beaked; achenes lenticular, 1.0–1.5 mm long, pale brown, iridescent; stigmas 2. May–June.

33. *Carex haydenii.*

a. Habit.
b. Pistillate scale.

c. Perigynium.
d. Achene.

Along streams, ditches, marshes, wet meadows, sometimes in shallow water. IA, IL, IN, MO, OH (OBL), NE (FACW+).

Hayden's sedge.

This species closely resembles *C. aquatilis* var. *substricta*, *C. emoryi*, and *C. stricta*. It differs from *C. aquatilis* var. *substricta* by its bladeless lower sheaths. It differs from *C. emoryi* and *C. stricta* in its nearly orbicular, nerveless perigynia that are exceeded by the pistillate scales.

32. **Carex hormathodes** Fern. Rhodora 8:165. 1906. Fig. 34.

Plants perennial, densely cespitose, from short, slender rhizomes; culms to 90 cm tall, somewhat 3-angled, scabrous only beneath the inflorescence; leaves 2–6 per culm, very slender, 1.0–2.5 mm wide, scabrous only at the tip; lowest sheaths closed, conspicuously nerved; inflorescence moniliform, arching to ascending, with 3–12 spikes, none of them touching the others; spikes gynecandrous, ovoid, tapering to either end, deep brown to stramineous, to 15 mm long, to 9 mm thick, with 20–35 perigynia; staminate and pistillate scales narrowly lanceolate, acuminate to short-aristate at the tip, nearly as long as the perigynia but much narrower; perigynia lance-ovoid, tawny, 4.5–6.0 mm long, 2.5–4.0 mm wide, broadest near the middle, tapering to a narrow beak, several-nerved on both faces; stigmas 2; achenes lenticular, 3–5 mm long. May–August.

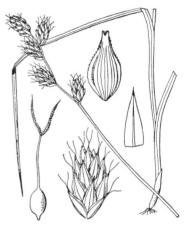

34. *Carex hormathodes*. Habit, perigynium, scale, seed, and spikelet.

Marshes, usually not in standing water.

IN (not listed for Region 3). This species is not indicated by the U.S. Fish and Wildlife Service for Indiana, but it is OBL elsewhere.

Marsh sedge.

This species is usually found in the maritime islands off the coast of Atlantic Canada and in Newfoundland and Quebec, then south to Virginia. The single location in Tippecanoe County, Indiana, is several hundred miles from its nearest station.

Carex hormathodes differs from other species of *Carex* in the Ovales group by its rather broad perigynia, its awn-tipped scales, and its spikes that taper to either end.

33. **Carex hyalinolepis** Steud. Syn. Cyp. 235. 1855. Fig. 35.
Carex lacustris Willd. var. *laxiflora* Dewey, Am. Journ. Sci. II, 35:60. 1863.
Carex riparia Curtis var. *impressa* S. H. Wright, Bull. Torrey Club 9:151. 1882.
Carex impressa (S. H. Wright) Mack. Bull. Torrey Club 37:236. 1910.

Plants perennial, cespitose, with elongated rhizomes; culms to 1.2 m tall, stout, triangular, rough to the touch, at least beneath the inflorescence, brown at the base; sterile shoots usually present; leaves up to 1.5 cm wide, glaucous, sometimes septate-nodulose, scabrous along the margins; lower sheaths white to pale brown,

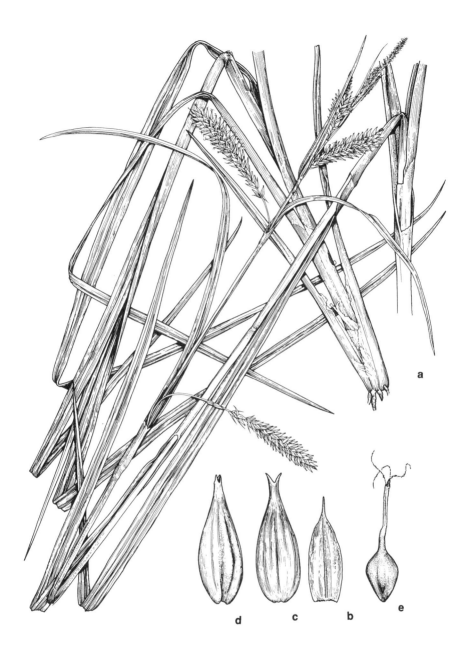

35. *Carex hyalinolepis.*

a. Habit.
b. Pistillate scale.

c, d. Perigynia.
e. Achene.

blade-bearing, rarely fibrillose, the ligule usually wider than long; upper 2–5 spikes staminate, up to 4 cm long, up to 5 mm thick, sessile except for the uppermost; staminate scales acute to short-awned, reddish brown with hyaline margins; pistillate spikes 2–4, appressed, up to 7.5 cm long, up to 1.5 cm thick; pistillate scales lance-ovate, acute to awned, reddish brown with hyaline margins, shorter than the perigynia; perigynia up to 150 per spike, ovoid to broadly ellipsoid, 6–8 mm long, 2.0–2.5 mm broad, ascending, coriaceous, glabrous, dull green, finely nerved, tapering to a bidentate beak up to 1 mm long; achenes trigonous, 2.0–2.5 mm long, black; stigmas 3. April–July.

Ditches, swamps, sloughs, fens, marshes, along streams, around ponds.

IA, IL, IN, KS, KY, MO, NE, OH (OBL).

Ditch sedge.

This species is very similar in appearance to *C. lacustris,* but it differs in its blade-bearing, pale brown lower sheaths and its more finely nerved perigynia. The blade-bearing, lower sheaths in *C. lacustris* are reddish or purple.

34. **Carex hystericina** Muhl. ex Willd. Sp. Pl. 4:282. 1805. Fig. 36.

Plants perennial, densely cespitose, from short, stout rhizomes; culms to 1 m tall, stout, triangular, scabrous, at least beneath the inflorescence, purplish at the base, with last year's leaves persistent; sterile shoots present; leaves up to 10 mm wide, firm, scabrous on both surfaces, green, septate-nodulose, at least some of them overtopping the inflorescence; lower sheaths red, the lowest bladeless, the upper pale, greenish, septate-nodulose, the ligule as wide as long; terminal spike 1 or rarely 2, staminate, to 5 cm long, to 4 mm thick, pedunculate, usually subtended by a leaflike bract; staminate scales awned, reddish brown with a green center; lateral spikes 1–4, pistillate, elongate-cylindric, 2–5 cm long, 1.0–1.5 cm thick, with capillary, scabrous peduncles, the lowest spreading to pendulous; bracts leaflike; pistillate scales ovate, with long, serrulate awns, reddish brown with a green center, usually reaching at least to the base of the perigynium; perigynia up to 100 per spike, crowded, lance-ovoid to ovoid, 4–6 mm long, 1.5–2.0 mm thick, spreading but not reflexed, inflated, pale green to stramineous, strongly 15- to 20-nerved, abruptly tapering to a stout bidentate beak 1.8–2.2 mm long, the teeth of the beak 0.2–0.7 mm long; achenes 1.5–1.8 mm long, trigonous, brown, obovoid, continuous with the persistent, flexuous style; stigmas 3, reddish brown. May–July.

Swamps, fens, ditches, along streams, sometimes in standing water.

IA, IL, IN, KS, KY, MO, NE, OH (OBL).

Porcupine sedge.

Carex hystericina differs from the similar *C. comosa* in that none of the perigynia is reflexed. It differs from the somewhat similar *C. lurida* and *C. baileyi* in its narrower, lance-ovoid perigynia, its more numerous strong nerves on its perigynia, and its obovoid achenes. Most botanists in the past have spelled the epithet *hystricina*.

35. **Carex interior** Bailey, Bull. Torrey Club 20:426. 1893. Fig. 37.

Plants perennial, densely cespitose, from short rhizomes; culms triangular, smooth or scabrous only below the head, to 90 cm tall; leaves 3–5, 0.5–2.5 mm wide, flat, green, smooth or slightly scabrous along the margins, shorter than the

36. *Carex hystericina*.

a. Habit.
b. Pistillate scale.
c. Perigynium.
d. Achene.

37. *Carex interior.*
a. Habit.
b. Inflorescence.

c. Pistillate scale.
d. Perigynium, dorsal view.
e. Perigynium, ventral view.

f. Achene.

culms; sheaths tight, smooth, the inner band hyaline and sometimes purple-dotted, concave at the apex, at least the lower ones brown to stramineous; inflorescence up to 3.5 cm long, with 2–5 sessile, separated spikes; terminal spike clavate, 5–20 mm long, gynecandrous, the staminate part 2–14 mm long, 3- to 10-flowered, the pistillate part 3–7 mm long, 4- to 16-flowered; lateral spikes pistillate or with a few staminate flowers at the base, 3–10 mm long, with scalelike bracts; pistillate scales ovate, 1.0–2.5 mm long, castaneous with a green center and hyaline margins, acute or obtuse at the tip, sometimes reaching the base of the beak of the perigynium; perigynia 2–3 mm long, 1.0–1.8 mm wide, ovoid to broadly ovoid, plano-convex, the ventral surface nerveless or with a few short nerves, spongy-thickened at the base, green to dark brown, sessile, tapering to a beak with at least the lower perigynia spreading to reflexed; beak of the perigynium 0.5–1.0 mm long, serrulate, toothed at the apex; achenes biconvex, about 1.5 mm long, up to 1.5 mm broad, substipitate, jointed to the deciduous style; stigmas 2. April–June.

Fens, swamps, marshes, bogs, wet meadows, moist prairies, wet woods.

IA, IL, IN, KS, KY, MO, NE, OH (OBL).

Inland sedge.

This small species is distinguished by its star-shaped spikes due to the radiating perigynia, its elongated staminate part of the terminal spike, its narrow leaves, and its often nerveless ventral face of the perigynium. The similar *C. atlantica* and *C. echinata* differ in their nerved perigynia. *Carex seorsa*, which is also similar, has a smooth, rather than serrulate beak on the perigynium.

36. **Carex joori** L. H. Bailey, Proc. Am. Acad. Arts & Sci. 22 (1):72. 1886. Fig. 38.

Plants perennial, cespitose, from short, rather thick rhizomes; culms up to 1.3 m tall, stout, strongly 3-angled, more or less glaucous, scabrous at least in the upper part; leaves 5–10 per culm, up to 10 mm wide, glaucous, stiff, long-tapering to the tip, the margins scabrous and revolute; lowest sheaths reddish brown to black, strongly fibrillose; ligules wider than long; terminal spike staminate, 1.5–5.0 cm long, rarely with 1–2 smaller staminate spikes as well, on scabrous peduncles up to 5 cm long; staminate scales elliptic to narrowly ovate, 4–7 mm long, tapering to an awn; pistillate spikes 3–5, cylindric, up to 6 cm long, up to 1 cm thick, on short, erect peduncles; pistillate scales oblong to elliptic to obovate, obtuse but awned at the tip, 3–6 mm long,

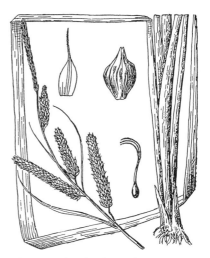

38. *Carex joori*. Habit, scale, perigynium, and achene.

reddish brown with a green center, shorter than the perigynia; perigynia squarrose, ascending, glaucous or dull green, obovoid, tapering to a minute beak, inflated, 3.5–5.0 mm long, prominently several-nerved; stigmas 3; achenes strongly trigonous, obovoid, 2.2–2.8 mm long, yellow-brown. August–October.

Swamps, bottomland forests, sometimes in shallow water.

MO (OBL).

Joor's sedge.

This southern species and *C. glaucescens* are very similar in appearance. Both are usually glaucous, and their perigynia closely resemble each other. In *C. joori*, all the pistillate spikes are erect to ascending, while in *C. glaucescens*, the lower pistillate spikes are pendulous.

37. **Carex lacustris** Willd. Sp. Pl. 4:306. 1805. Fig. 39.

Plants perennial, cespitose, with elongated rhizomes; culms to 1.3 m tall, stout, triangular, rough to the touch, at least beneath the inflorescence, usually purplish at the base; sterile shoots usually present; leaves up to 1.5 cm wide, glaucous at least below, firm, septate-nodulose, scabrous along the margins; lower sheaths red to purplish, bladeless, becoming fibrillose, the ligule much longer than wide; upper 2–5 spikes staminate, to 5 cm long, to 4 mm thick, sessile except for the uppermost; staminate scales obtuse to awned, reddish with usually hyaline margins; pistillate spikes 2–3, appressed, on short stalks, to 8 cm long, to 1.5 cm thick; pistillate scales lance-ovate, acute to acuminate, awned, reddish with hyaline margins and a green center, shorter than the perigynia; perigynia up to 150 per spike, lance-ovoid, 5–7 mm long, about 2 mm broad, ascending, coriaceous, glabrous, olive green, strongly nerved, tapering to a bidentate beak up to 1.5 mm long; achenes trigonous, 2.2–2.5 mm long, black; stigmas 3. May–July.

Around lakes, marshes, swamps, wet meadows, ditches.

IA, IL, IN, KS, MO, NE, OH (OBL).

Lake sedge.

This stout species is very similar to *C. hyalinolepis*, but differs in its reddish or purple, bladeless, lower sheaths and its strongly nerved perigynia.

38. **Carex laeviconica** Dewey, Am. Journ. Sci. II, 24:47. 1857. Fig. 40.

Plants perennial, cespitose, with slender creeping rhizomes; culms up to 1.2 m tall, stout, triangular, usually somewhat scabrous beneath the inflorescence, purplish at the base; sterile shoots usually present; leaves up to 8 mm wide, dull green, glabrous, septate-nodulose, scabrous along the margins; sheaths tight, brown, the middle and upper ones cartilaginous, globose or slightly scabrous, the lower reddish, becoming fibrillose, the ligule as long as wide; upper 2–4 spikes staminate, up to 5 cm long, up to 4 mm thick, all but the uppermost sessile; staminate scales obtuse, usually awned, yellow-brown with hyaline margins; pistillate spikes 2–4, ascending, up to 7.5 cm long, up to 1.2 cm thick; bracts leaflike; pistillate scales ovate, acute and often awned, reddish brown with a hyaline margin, shorter than the perigynia; perigynia up to 50 per spike, ovoid, 5–7 mm long, ascending, coriaceous, yellow-green to stramineous, strongly nerved, tapering to a smooth or serrulate bidentate beak up to 1.6–2.0 mm long, the teeth 1–2 mm long; achenes trigonous, 2.2–2.5 mm long, brown; stigmas 3. April–July.

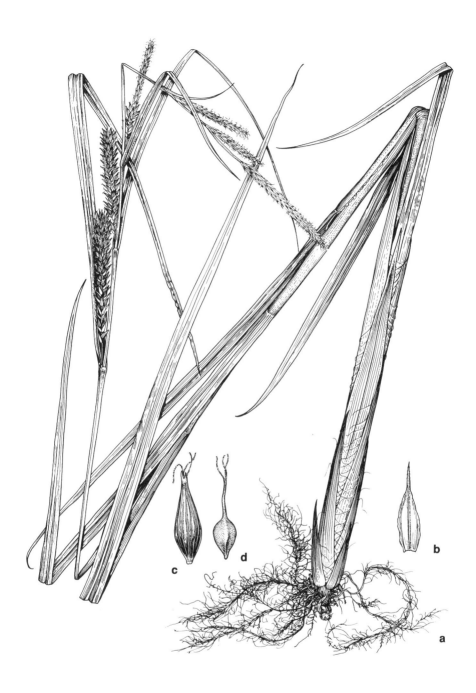

39. *Carex lacustris.*

a. Habit.
b. Pistillate scale.
c. Perigynium.
d. Achene.

40. *Carex laeviconica.*

a. Habit.
b. Pistillate scale.
c. Perigynium.
d. Achene.

Sloughs, marshes, along streams, around lakes and ponds, swamps, wet meadows, ditches.

IA, IL, IN, KS, MO, NE (OBL).

Smoothbeak sedge.

This species is distinguished by the teeth of the perigynium being 1–2 mm long, the leaves being no more than 8 mm wide, and the ovoid perigynia being 5–7 mm long.

39. **Carex laevivaginata** (Kukenth.) Mack. in Britt. & Brown, Ill. Fl., ed. 2, 1:371. 1913. Fig. 41.
Carex stipata Muhl. var. *laevivaginata* Kukenth. in Engler, Pflanzenreich 4 (20):172. 1909.

Plants perennial, densely cespitose, with fibrous roots and short, stout, black rhizomes; culms soft-spongy in fresh plants, easily compressed, stout, erect, to 80 cm tall, strongly 3-angled, with the sides concave, rough to the touch on the angles, usually overtopped by the leaves; leaves 3–6, up to 60 cm long, 3–6 mm wide, flat, flaccid, pale green, rough on the margins and usually on the midvein beneath; sheaths open, loose, pale green to yellow, strongly nerved, concave to truncate and cartilaginous at the mouth, not breaking easily, usually septate-nodulose, white-hyaline ventrally; ligule longer than wide; inflorescence a compound, elongated, usually continuous head, up to 6 cm long, up to 1.5 cm broad; bracts setaceous to scalelike, rough-awned, up to 4 cm long, seldom longer than the spike it subtends; spikes 10–15, the staminate flowers above the pistillate; pistillate scales lanceolate, acuminate to long-awned, hyaline with 3 green nerves, three-fourths as long as and narrower than the perigynia; staminate scales narrowly lanceolate, pale brown; perigynia 4–10 per spike, lanceoloid, 4.5–8.0 mm long, 1.25–2.25 mm broad, plano-convex, rounded and spongy thickened at base, stipitate, greenish or yellowish, conspicuously 2- to 12-nerved, tapering gradually to a 2-toothed, serrulate beak 2.0–2.5 mm long, the teeth appressed to slightly spreading; achenes 1.5–2.0 mm long, 1.25 mm wide, lenticular, ovoid to suborbicular, apiculate, substipitate, jointed to the style; stigmas 2, reddish brown, slender, flexuous. May–June.

Along streams, ditches, fens, swamps, marshes, wet meadows.

IA, IL, IN, KY, MO, OH (OBL).

Soft sedge.

This species is very similar to the much more common *C. stipata*, from which it differs in its leaf sheaths that are cartilaginous near their summit; the shorter, less-branched inflorescence; and the usually green perigynia.

40. **Carex lasiocarpa** Ehrh. var. **americana** Fern. Rhodora 44:304. 1942. Fig. 42.

Plants perennial, solitary or in small tufts, with long, scaly stolons; sterile shoots common; culms to 1.2 m tall, wiry, usually smooth, purple-red at the base; leaves up to 2 mm wide, convolute, septate between the veins, but with no midvein, usually roughened near the tip; lowest sheaths reddish, becoming fibrillose; upper two spikes usually staminate, erect, to 6 cm long, on slender peduncles; staminate scales light red-brown; lowest 1–3 spikes pistillate, sometimes with a few staminate flowers at the tip, to 5 cm long, erect, sessile or subsessile; pistillate scales lanceolate, acute to awned, purple-brown with a green center, usually slightly shorter than the

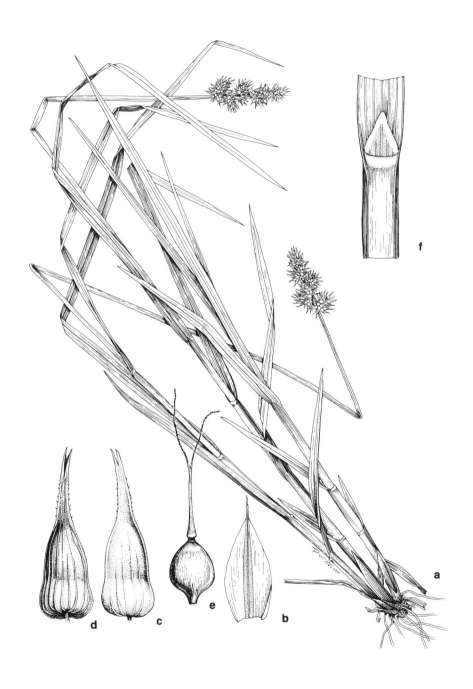

41. *Carex laevivaginata.*
a. Habit.
b. Pistillate scale.

c. Perigynium, dorsal view.
d. Perigynium, ventral view.
e. Achene.

f. Sheath with ligule.

42. *Carex lasiocarpa*
var. *americana.*

a. Habit.
b. Pistillate scale.

c. Perigynium.
d. Achene.

perigynia; perigynia up to 50 per spike, broadly ovoid to ellipsoid, 3–5 mm long, densely hairy, obscurely nerved, with a short, bidentate beak, the teeth of the beak up to 0.6 mm long; achenes trigonous, 1.7–1.9 mm long, yellow-brown, punctate, jointed with the style; stigmas 3. May–June.

Swamps, sloughs, around lakes and ponds.

IA, IL, IN, MO, OH (OBL).

Hairy-fruited sedge.

This plant is readily distinguished by its very narrow, convolute leaves that lack a midvein, by its smooth culms, and by its hairy perigynia.

41. **Carex leptalea** Wahl. Sv. Vet. Akad. 24:139. 1803. Fig. 43.
Carex polytrichoides Willd. Sp. Pl. 4:213. 1805.

Plants perennial, densely cespitose, with slender, scaly stolons; culms to 60 cm tall, capillary, weak, sharply triangular, longer than the leaves, brown at base, with old leaf bases persisting; leaves usually 2 per culm, filiform, up to 1.3 mm wide, flat or canaliculate, deep green, smooth or scabrous on the margins, minutely papillose on the upper surface; sheaths tight, the ventral band hyaline, sometimes strongly nerved and becoming fibrillose, truncate or concave at summit, the lowest pale or brown; spike 1 per culm, terminal, up to 15 cm long, up to 3 mm wide, ovoid to oblongoid, androgynous with a few appressed staminate scales at the tip; bracts absent; pistillate scales ovate, obtuse to acute to short-awned, about 1/2 as long as the perigynia, with a green center and hyaline margins; perigynia 1–10 per spike, 2.5–4.0 mm long, ellipsoid, strongly ascending, flattened and 2-edged, pale green to yellowish, several-nerved, more or less spongy at base, substipitate, beakless; achenes trigonous, 2.0–2.5 mm long, 1.5–2.0 mm wide, shiny; stigmas 3, short, reddish brown. May–July.

Along streams, fens, wet woods.

IA, IL, IN, KY, MO, OH (OBL).

Slender sedge.

This is the only *Carex* with capillary culms, filiform leaves, solitary spikes with a few male flowers at the tip, and beakless perigynia that are several-nerved.

42. **Carex limosa** L. Sp. Pl. 1:977. 1753. Fig. 44.

Plants perennial, with long-running, slender, brown, scaly stolons; culms to 60 cm tall, slender, scabrous on the sharp angles, purple-red at the base; leaves 1–3 near the base of the plant, up to 2.5 mm wide, conduplicate to involute, keeled, glaucous, scabrous along the margins; sheaths reddish, the hyaline ventral band russet-spotted and pale-nerved near the apex; terminal spike staminate, to 3.5 cm long, on long, slender peduncles; lateral spikes 1–3, pistillate, sometimes with a few staminate flowers at the tip, up to 2 cm long, 4–8 mm thick, on slender, ascending peduncles; pistillate scales ovate, obtuse to acuminate, often mucronate, yellow-brown to purple-brown, with a green midvein, about as long as or slightly longer than the perigynia; perigynia up to 30 per spike, 3–4 mm long, ovoid, trigonous but compressed, glaucous green, papillose, several-nerved, substipitate, minutely beaked; achenes trigonous, 2.0–2.3 mm long, dark brown, apiculate; stigmas 3. June–July.

43. _Carex leptalea._
a. Habit.

b. Inflorescence.
c. Scale.

d. Perigynium.
e. Achene.

44. *Carex limosa*.

a. Habit.
b. Pistillate scale.
c. Perigynium.
d. Achene.

Bogs, marshes.

IA, IL, IN, KY, MO, NE, OH (OBL).

Pendulous bog sedge.

This slender species is recognized by its narrow, often involute leaves, its long-pedunculate staminate spike, and its ovoid, glaucous perigynia.

43. **Carex longii** Mack. Bull. Torrey Club 49:372. 1923. Fig. 45.

Plants perennial, densely cespitose, from short, black, fibrillose rootstocks; culms to 1.2 m tall, wiry, triangular, scabrous on the angles beneath the inflorescence or smooth, longer than the leaves, pale brown to brownish black at the base; leaves 2–4 per culm, ascending, 2–5 mm wide, flat, the upper surface minutely papillose, the veins on the lower surface and the margins usually scabrous; sheaths somewhat loose, green-nerved throughout, hyaline ventrally, prolonged at the summit, the lowest ones stramineous; spikes 3–10 per culm, 6–13 mm long, not strongly clavate at base, green to brownish, approximate or somewhat separated but rarely moniliform, the inflorescence 1.0–4.5 cm long, gynecandrous; lowest bract usually setaceous, scabrous, the upper bracts scalelike; pistillate scales lanceolate, obtuse to acute, concave, shorter than the perigynia, the center greenish and 3-nerved, the nerves not quite reaching the tip of the scale, the margins silver-hyaline; perigynia many per spike, 3.0–4.5 mm long, 1.6–2.6 mm wide, obovate, widest near the middle, appressed-ascending, plano-convex, green to brownish, papery, strongly nerved on the outer face, with 4–7 raised nerves on the inner face, winged all the way to the tip as well as to the base, substipitate, the beak 0.5–1.0 mm long, serrulate, bidentate; achenes lenticular, 1.3–1.7 mm long, 0.7–1.0 mm wide, apiculate, stipitate, weakly jointed with the deciduous style; stigmas 2, short, reddish. April–June.

Wet woods, wet sand prairies, flatwoods, sometimes in pools of shallow water.

IA, IL, IN, KS, KY, MO, OH (OBL).

Long's sedge.

This species is similar to and often confused with *C. albolutescens*. It differs from *C. albolutescens* in its appressed-ascending perigynia that are winged from the base all the way to the tip of the beak and in its convex pistillate scales whose midvein fails to reach the tip of the scale.

44. **Carex lupuliformis** Sartw. ex Dewey, Am. Journ. Sci. II, 9:29. 1850. Fig. 46.
Carex lupulina Muhl. var. *polystachia* Schw. & Torr. Ann. Lyc. N.Y. 1:337. 1825.
Carex lurida Wahl. var. *polystachya* (Schw. & Torr.) L. H. Bailey, Proc. Am. Acad. 22:63. 1886.
Carex eggertii L. H. Bailey, Bot. Gaz. 21:6. 1896.

Plants perennial, cespitose or growing singly, from fibrous roots, with long, slender, connected rhizomes; culms up to 1.06 m tall, stout, much exceeded by the upper leaves and bracts, triangular, smooth throughout, light brown or green, becoming purplish at base; leaves up to 60 cm long, 5.0–11.5 mm wide, tapering gradually to an attenuated tip, septate-nodulose, flat, dull green, with rough margins; sheaths short prolonged at mouth, white-hyaline ventrally; staminate spikes 1–3, linear, 2.0–10.5 cm long, 2–6 mm broad, on short or long peduncles,

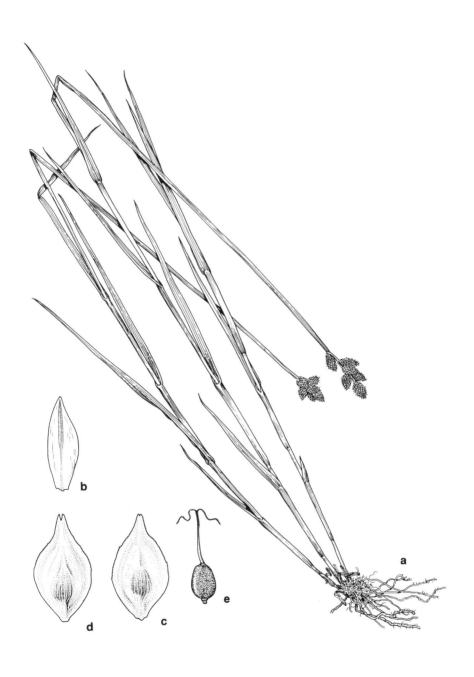

45. *Carex longii.*
a. Habit.

b. Pistillate scale.
c. Perigynium, dorsal view.

d. Perigynium, ventral view.
e. Achene.

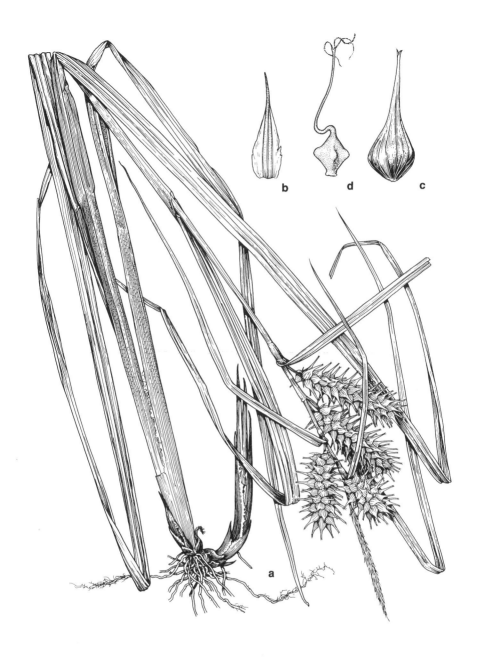

46. *Carex lupuliformis.*

a. Habit.
b. Pistillate scale.

c. Perigynium.
d. Achene.

the scales linear to lanceolate to ovate, strongly awned to acuminate, stramineous with a several-nerved green center with hyaline margins; bracts shorter to much exceeding the spikes; pistillate spikes 2–7, oblong-cylindric, 1.5–9.6 cm long, 1.5–3.5 cm broad, erect, densely flowered, sessile or on peduncles up to 9 cm long; bracts leaflike, exceeding the inflorescence, with sheaths strongly prolonged at the mouth; perigynia 10–87, diverging from the axis of the pistillate spike at an angle of 30–90 degrees; pistillate scales lanceolate, strongly awned to acute, stramine-ous, with a several-nerved green center and hyaline margins, much narrower than and usually shorter than the perigynia; perigynia subuloid, 10–19 mm long, 3.0–5.5 mm broad, strongly inflated, glabrous, dull green or brownish yellow at maturity, sessile, rounded at the base, with 18–30 nerves, tapering to a beak 1/2 or slightly less than 1/2 the length of the total perigynia, the beak bidentate, with teeth up to 2 mm long, smooth within; achenes rhomboid to obtrulloid, wider than long to longer than wide, 2–4 mm long, 2.0–3.5 mm broad, the sides usually concave, occasionally flat to slightly convex, with the angles acute, upturned, or with well-defined knobs, tapering to the base, tapering into and continuous with the persis-tent abruptly bent or fully coiled style; stigmas 3, short, slender, blackish. July–September.

Swamps, around ponds, sloughs.

IA, IL, IN, KY, MO, OH (FACW+).

Hoplike sedge.

Carex lupuliformis and *C. lupulina* are very similar in appearance, although *C. lupuliformis* usually has larger pistillate spikes. The most reliable character to distinguish the two is that *C. lupuliformis* has achenes that have the angles knobbed or acute and upturned and with usually concave sides.

In the southern part of the central Midwest, there are specimens that have achenes that appear to be 6-sided. These have been called *C. eggertii*.

45. **Carex lupulina** Willd. Sp. Pl. 4:266. 1805. Fig. 47.
Carex lupulina Willd. var. *pedunculata* Gray in Beck, Bot. U.S. 438. 1833.
Carex canadensis Dewey, Am. Journ. Sci. II, 41:229. 1866.
Carex lupulina Willd. var. *androgyna* Wood, Bot. & Fl. 376. 1870.
Carex lupulina Willd. var. *longipedunculata* Sart. ex Dudley, Bull. Cornell Univ. 2:119. 1886.
Carex gigantea Rudge var. *lupulina* (Willd.) Farw. Ann. Rfep. Comm. Parks Detroit II:50. 1900.
Carex lupulina Willd. var. *albomarginata* Sherff, Bull. Torrey Club 38:482. 1911.

Plants perennial, cespitose, from fibrous roots, with long connected rhizomes; culms up to 1.26 m tall, stout, much exceeded by the upper leaves and bracts, triangular, smooth throughout, light brown or green, becoming purplish near the base; leaves up to 60 cm long, 4.5–8.5 mm wide, tapering gradually to an attenu-ated tip, septate-nodulose, flat, dull green, with the margins rough; sheaths short-prolonged at the mouth, white-hyaline ventrally, the ligule usually longer than wide; staminate spike 1, rarely 2–3, linear, 2–9 cm long, 2.0–4.5 mm broad, nearly sessile to long peduncled; bracts somewhat shorter than to much exceeding the spike; pistillate spikes 1–6, oblongoid to oblongoid-cylindric, 1.9–7.0 cm long, 1.2–

a

47. _Carex lupulina._

a. Habit.
b. Pistillate scale.

c. Perigynium.
d. Achene.

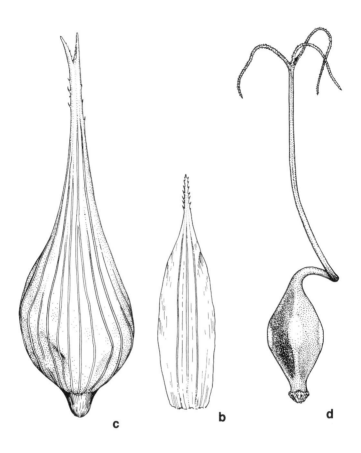

c   b   d

3.0 cm broad, erect, sessile or with peduncles up to 17.2 cm long, densely flowered; bracts leaflike, exceeding the inflorescence, sheathing; staminate scales linear to lanceolate, acuminate to aristate, stramineous, with hyaline margins and a strongly nerved green center; perigynia diverging from the axis of the pistillate spike at an angle of 30–60 degrees; pistillate scales lanceolate, rough-awned or acuminate, straw-colored with a several-nerved green center and hyaline margins, much narrower than and usually shorter than the perigynia; perigynia subuloid, 10–17 mm long, 2.5–5.0 mm broad, rounded at the base, strongly inflated, glabrous, green or brownish yellow at maturity, sessile or short-stipitate, with 18–22 nerves, tapering to a beak 1/2 or slightly less than 1/2 the total length of the perigynia, the beak bidentate, with the teeth up to 2 mm long, smooth within; achenes rhomboid to trulloid, longer than wide to as long as wide, 2.0–4.5 mm long, 1.25–3.00 mm broad, the sides usually concave, occasionally flat, with obtuse to acute angles and a slight suggestion of a knob, tapering to the base, tapering into and continuous with the persistent abruptly bent or fully coiled style; stigmas 3, short, slender, blackish. May–September.

Swamps, sloughs, marshes, ditches, along streams, around lakes and ponds. IA, IL, IN, KY, MO, OH (OBL), KS, NE (FACW+).

Hop sedge.

This species is distinguished by its large pistillate spikes that are longer than they are broad, by its large, beaked, crowded perigynia, and the achenes that usually have blunt angles and slightly concave to flat sides.

Carex lupulina is somewhat variable, attested to by the number of named variations listed in the taxonomy above.

46. **Carex lurida** Wahl. Sv. Vet. Akad. 24:153. 1803. Fig. 48.
Carex tentaculata Muhl. ex Willd. Sp. Pl. 4:266. 1805.

Plants perennial, densely cespitose, from short, stout rhizomes; culms to 1 m tall, stout, triangular, smooth or scabrous, purplish at the base, often with last year's leaves persistent; leaves up to 7 mm wide, glabrous, septate-nodulose, dull green, scabrous along the margins, some of them sometimes overtopping the culms; sheaths tight, pale or tan or the lower reddish, the ventral band weakly nerved, concave or truncate at the mouth, the ligule longer than wide; terminal spike 1, staminate, up to 7 cm long, up to 3 mm thick, sessile or short-pedunculate; staminate scales serrulate-awned, pale brown with hyaline margins; lateral spikes 1–4, pistillate, thick-cylindric to ovoid, 1.5–4.0 cm long, 1.5–2.0 cm thick, all but sometimes the lowest erect or ascending, on short, glabrous peduncles; bracts leaflike; pistillate scales linear, abruptly contracted to a serrulate awn, usually reaching the base of the beak of the perigynium; perigynia up to 100 or more per spike, crowded, 6–9 mm long, 3–4 mm thick, yellow-brown, membranous, inflated, shiny, glabrous, strongly nerved, spreading to ascending but not reflexed, gradually tapering to a bidentate beak 3–4 mm long, the beak about as long as the body; achenes trigonous, 2.0–2.5 mm long, yellow-brown, substipitate, continuous with the persistent style; stigmas 3. May–September.

Marshes, fens, along rivers and streams, around ponds and lakes, sinkhole ponds, wet meadows, swamps.

IA, IL, IN, KY, MO, OH (OBL).

Lurid sedge.

This species most nearly resembles *C. baileyi* from which it differs in its more tapering perigynial beak and its thicker pistillate spikes. It differs from *C. comosa* in that none of its perigynia is reflexed. From *C. hystericina*, it differs by its ovoid, thicker, inflated perigynia.

47. **Carex molesta** Mack. ex Bright, Trillia 9:30. 1930. Fig. 49.

Plants perennial, cespitose, from short, black, fibrillose rootstocks; culms to 1 m tall, wiry, triangular, the angles scabrous beneath the inflorescence, with old leaf bases often persisting; sterile shoots usually not present; leaves 4–7 per culm, ascending, 2.0–3.5 mm wide, flat, rather thin, light green, scabrous along the margins; sheaths tight, the ventral band narrowed quickly below the summit, pale green, prolonged at the summit, the lowermost brownish; spikes 2–5 per culm, 6–10 mm long, nearly globose, not clavate at base, green to greenish white, crowded into a dense inflorescence up to 3 cm long, or sometimes with the lower spikes distant; bracts mostly scalelike; pistillate scales lanceolate, obtuse to acute, barely reaching the base of the beak of the perigynium, the center green and 3-nerved, the margins hyaline; perigynia 15–30 per spike, 4.0–5.5 mm long, 2–3 mm wide,

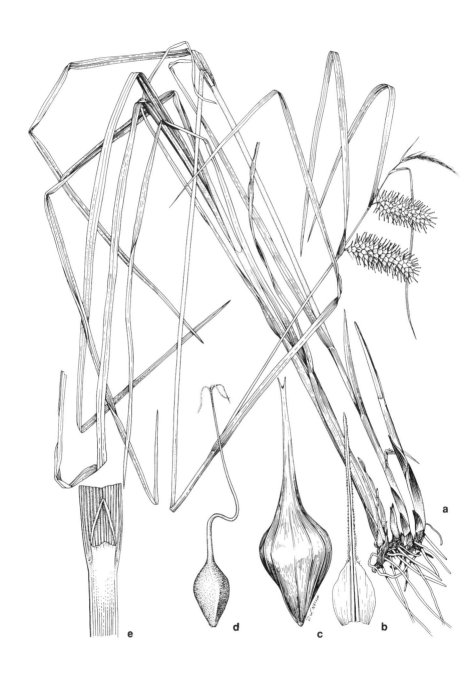

48. *Carex lurida.*
a. Habit.

b. Pistillate scale.
c. Perigynium.

d. Achene.
e. Sheath with ligule.

49. *Carex molesta.*
a. Habit.

b. Pistillate scale.
c. Perigynium, dorsal view.

d. Perigynium, ventral view.
e. Achene.

broadly ovate to suborbicular, widest near the middle, submembranous, ascending, plano-convex, green or greenish white, faintly nerved on the outer face, finely nerved on the inner face, winged to the base, substipitate, the beak 1.0–1.3 mm long, serrulate, bidentate, tipped with yellow-brown; achenes lenticular, 1.6–1.9 mm long, 1.2–1.3 mm wide, yellow-brown, apiculate, substipitate; stigmas 2, short, reddish brown. April–June.

Swamps, ditches, wet depressions in woods, wet prairies, rarely in standing water.

IA, IL, IN, MO (FAC+), KS, NE (FAC), KY, OH (FACU).

Confusing sedge.

The globose spikelets of *C. molesta* resemble those of *C. cristatella*, but all of the perigynia of *C. molesta* are ascending. *Carex molesta* also resembles some specimens of *C. brevior*, but it generally does not have quite orbicular perigynia, and the inner face of each perigynium is finely nerved.

48. **Carex muskingumensis** Schwein. Ann. Lyc. N.Y. 1:66. 1824. Fig. 50.

Plants perennial, densely cespitose, from short, stout, black, fibrillose rhizomes; culms to 85 cm tall, sharply triangular, the angles very scabrous above, stiffly erect, usually longer than the leaves, with many very leafy sterile shoots present; leaves up to 12 per fertile culm, oriented at right angles to the culm, 2.5–4.0 mm wide (leaves of sterile culm up to 7 mm wide), flat, the margins scabrous; sheaths tight, striate ventrally to the V-shaped mouth, the mouth with a thickened, dark band, the ligule short, rigid, and holding the leaf base away from the culm; spikes 6–10 per culm, 12–25 mm long, narrowly ellipsoid, light green to stramineous, appressed-ascending, acuminate, gynecandrous, long-clavate at the base, overlapping to aggregated, forming an inflorescence 4.5–9.0 cm long, 10–20 mm wide; bracts leaflike, with prolonged, rough awns below, awnless above, rarely longer than the adjacent spike; pistillate scales lanceolate, acuminate to obtuse, two-thirds as long as and much narrower than the perigynia, tan to brown, hyaline with a darkened center; perigynia many per spike, lanceoloid to narrowly ovoid-lanceoloid, plano-convex, appressed, 7–10 mm long, 1.75–2.00 mm wide, broadest at or above the middle, substipitate, distinctly nerved dorsally and ventrally, the wing abruptly narrowed at or just below the middle, absent at the base, the beak 4.0–4.8 mm long, serrulate, sharply bidentate; achenes lenticular, 2.5–3.0 mm long, about 0.75 mm wide, apiculate, stipitate, reddish brown, jointed to and sometimes continuous with an elongated style; stigmas 2, thin, flexuous, reddish. May–June.

Low, swampy woods, often in wet depressions, floodplains.

IA, IL, IN, KY, MO, OH (OBL), KS, NE (NI).

Muskingum sedge.

The large, pointed, gynecandrous spikes, the very narrow perigynia, and the leaves that often stand at right angles to the culm and in perfect three ranks are distinctive for this species.

49. **Carex normalis** Mack. Bull. Torrey Club 37:244. 1910. Fig. 51.
Carex mirabilis Dewey, Am. Journ. Sci. 30:63. 1836, non Host. (1809).
Carex mirabilis Dewey var. *perlonga* Fern. Proc. Am. Acad. 37:473. 1902.
Carex normalis Mack. var. *perlonga* (Fern.) Burnham, Torreya 19:131. 1919.
Carex normalis Mack. f. *perlonga* (Fern.) Fern. Rhodora 44:285. 1942.

50. *Carex muskingumensis.*
a. Habit.
b. Pistillate scale.

c. Perigynium, dorsal view.
d. Perigynium, ventral view.
e. Achene.

f. Sterile culm.
g. Sheath, dorsal view.
h. Sheath with ligule.

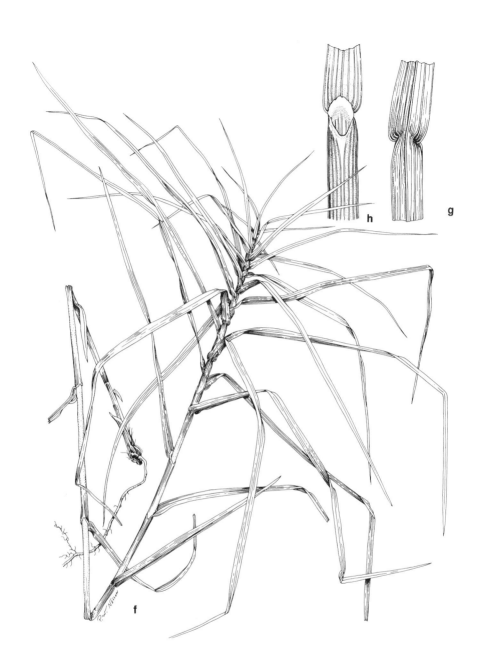

f

g

h

51. *Carex normalis.*
a. Habit.
b. Spike.
c. Pistillate scale.
d. Perigynium, dorsal view.
e. Perigynium, ventral view.
f. Achene.

Plants perennial, densely cespitose, from short, compact, brown, scaly to fibrillose rootstocks; culms to 1.5 m tall, wiry, sharply triangular, scabrous on the angles beneath the inflorescence, longer than the leaves, with several sterile culms; leaves 3–7 per culm, ascending, 2.0–6.5 mm wide, flat, lax, green, the margins scabrous toward the apex; sheaths loose, green-mottled, at least partially septate, prolonged and slightly concave at the mouth, the lowermost brownish; spikes 4–10 per culm, 6–10 mm long, green, gynecandrous, subglobose, clavate with the staminate flowers conspicuous, sometimes crowded into a head, sometimes separated in a moniliform inflorescence up to 5 cm long; lowest bracts setaceous, scabrous, upper bracts scalelike; pistillate scales ovate, obtuse to acute, barely reaching the base of the beak of the perigynium, the center green to brown and 3-nerved, the margins white- to tan-hyaline; perigynia many per spike, 3.0–4.5 mm long, 1.5–2.1 mm wide, ovate, widest at the middle, spreading, plano-convex, distinctly and finely nerved on both faces, narrowly and evenly winged to the base, substipitate, the beak 0.8–1.2 mm long, serrulate, bidentate; achenes lenticular, 1.6–1.8 mm long, 1.1–1.3 mm wide, apiculate, stipitate, weakly jointed with the deciduous style; stigmas 2, short, slender, reddish brown. May–June.

Seep springs, mesic woods, floodplains, streambanks, mesic savannas, marshes, borders of ponds, moist fields, ditches, marshes, not usually in standing water.

IA, IL, IN, MO (FACW), KS, NE (FACW-), KY, OH (FACU).

Sedge.

This species is distinguished by its numerous sterile culms, by its perigynia that are up to 2.1 mm wide and winged to the base, and by its spikes that are about as wide as they are long. There is considerable variation in leaf width, perigynia length, and arrangement of spikes in the inflorescence. Specimens with moniliform spikes may be called f. *perlonga*.

50. **Carex oligosperma** Michx. Fl. Bor. Am. 2:174. 1803. Fig. 52.

Plants perennial, loosely cespitose, with long, slender, deep rhizomes; culms to 1 m tall, somewhat triangular, firm, filiform, scabrous at least beneath the inflorescence, purple-red at the base, with last year's leaves persistent; leaves filiform, involute, to 40 cm long, to 3 mm wide, firm, septate-nodulose, scabrous along the margins; sheaths tight, green to pink-tinged, the ligule wider than long; terminal spike 1, staminate, to 4 cm long, to 1.5 mm thick, on a scabrous peduncle; staminate scales obtuse, yellow-brown with hyaline margins; lateral spikes 1–3, pistillate, short-cylindric to ovoid, to 2 cm long, to 1 cm thick, sessile; lowest bract leaflike; pistillate scales broadly ovate, acute to cuspidate, brown with hyaline margins and a green center, shorter than the perigynia; perigynia up to 15 per spike, ovoid, 4–7 mm long, 2.5–3.0 mm wide, inflated, subcoriaceous, shiny, yellow-green, strongly nerved, tapering to a smooth, bidentate beak 1–2 mm long; achenes trigonous, 2.8–3.0 mm long, about 2 mm wide, brown, substipitate, continuous with the style; stigmas 3. May–June.

Sphagnum bogs.

IL, IN, OH (OBL).

Few-seeded sedge.

Carex oligosperma is readily distinguished by its involute-filiform leaves, its inflated perigynia, its slender staminate spike, and its few, widely separated pistillate spikes.

52. *Carex oligosperma.*

a. Habit.
b. Pistillate scale.
c. Perigynium.
d. Achene.

51. **Carex pauciflora** Lightf. Fl. Scot. 2:543, t. 6, f. 2. 1771. Fig. 53.

Plants perennial, cespitose, from long rootstocks; culms to 60 cm tall, very slender but stiff, glabrous; leaves 2–3 per culm, shorter than the culm, to 2 mm wide, glabrous; lowest sheaths bladeless; inflorescence a solitary, terminal, androgynous spike; spike narrowly lanceoloid, up to 1 cm long, with 1–6 staminate and 1–6 pistillate flowers; scales lanceolate to ovate, caducous, about one-third as long as the perigynia; perigynia oblongoid, long-acuminate at the tip, yellow to stramineous, 6–8 mm long, 0.7–1.5 mm wide, obscurely nerved, strongly reflexed; stigmas 3; achenes linear to narrowly oblongoid, 4–6 mm long, smooth. June–July.

53. *Carex pauciflora.* Habit, perigynium with scale, and staminate flower.

Sphagnum bogs, tamarack bogs.

IN (OBL).

Few-flowered sedge.

This slender sedge is distinguished by its solitary, androgynous spike with 1–6 linear to narrowly oblongoid perigynia that are 6–8 mm long. The lowest sheaths do not bear blades.

52. **Carex paupercula** Michx. Fl. Bor. Am. 2:172. 1803. Fig. 54.
Carex paupercula Michx. var. *pallens* Fern. Rhodora 8:76–77. 1906.

Plants perennial, cespitose, from yellowish rootstocks; culms to 75 cm tall, slender, sharply 3-angled, scabrous; leaves 2–3 mm wide, flat, pale green, glabrous, shorter than the culms; sheaths often reddish, the hyaline ventral band usually pale-nerved near the summit; lowest bract leaflike, longer than the inflorescence; terminal spike staminate or sometimes gynecandrous, narrowly ellipsoid, to 1.5 cm long; staminate scales linear-lanceolate, acuminate; pistillate spikes 2–4, subglobose to oblongoid, to 18 mm long, to 8 mm thick, on pendulous, filiform peduncles up to 3 cm long, with up to 25 perigynia; pistillate scales lanceolate

54. *Carex paupercula.* Habit, achene, perigynium, and pistillate scale.

to narrowly ovate, long-acuminate to short-awned, 5–8 mm long, longer than the perigynia, pale brown to stramineous; perigynia broadly ovoid to suborbicular, glaucous to pale green, 3–4 mm long, 1.5–2.5 mm broad, few-nerved, papillate; stigmas 3; achenes trigonous, smooth, 2.5–3.5 mm long. June–August.

Wet meadows, bogs.

IN (OBL). This species is called *Carex magellanica* Lam. by the U.S. Fish and Wildlife Service.

Depauperate sedge.

This species is similar and closely related to *C. limosa*, differing by its flat, pale green leaves rather than involute, glaucous leaves. It also differs by its long-acuminate to short-awned pistillate scales.

Carex paupercula is a northern species that barely enters the central Midwest in northern Indiana. Our plants, which have pale scales, may be distinguished as var. *pallens*.

53. **Carex pellita** Willd. in Schk. Riedgr. 84. 1801. Fig. 55.
Carex lanuginosa Michx. Fl. Bor. Am. 2:175. 1803.
Carex filiformis L. var. *latifolia* Boeck. Linnaea 41:309. 1877.

Plants perennial, loosely cespitose, with long, scaly stolons; sterile shoots common; culms to 1 m tall, stiff, usually scabrous on the angles, sometimes smooth, purple-red at the base; leaves up to 5 mm wide, flat except for the revolute margins, septate, with a conspicuous midvein, scabrous toward the tip; lowest sheaths dark red, becoming fibrillose; upper 1–3 spikes staminate, erect, up to 4 cm long, long-pedunculate; staminate scales light reddish brown; lowest 2–3 spikes pistillate, occasionally with a few staminate flowers at the tip, up to 5 cm long, erect, sessile or subsessile; pistillate scales broadly lanceolate, acuminate to awned, red-brown with a green center, as long as or longer than the perigynia; perigynia up to 75 per spike, broadly ovoid, 2.0–3.5 mm long, densely pubescent, many-nerved, with a bidentate beak about 1 mm long, the teeth of the beak 0.4–0.8 mm long; achenes trigonous, 1.7–1.9 mm long, yellow-brown, punctate, jointed with the style; stigmas 3. April–July.

Marshes, fens, wet prairies, swamps.

IA, IL, IN, KS, KY, MO, NE, OH (OBL). This species is called *Carex lanuginosa* by the U.S. Fish and Wildlife Service.

Woolly sedge.

This species is readily distinguished by its flat blades up to 5 mm wide, its usually multiple staminate spikes, and its densely woolly-hairy perigynia. It differs from the similar *C. lasiocarpa* var. *americana* by its broader leaves with a distinct midvein.

54. **Carex prairea** Dewey ex Wood, Class-Book, ed. 2, 414. 1847. Fig. 56.
Carex teretiuscula Gooden var. *ramosa* Boott, Illustr. Carex 43. 1867.
Carex diandra Schrank var. *ramosa* (Boott) Fern. Rhodora 10:48. 1908.

Plants loosely tufted or more or less running; rhizomes short, slender, blackish, fibrous-scaly; roots wiry, abundant; culms 1–several, mostly fertile, erect, becoming slightly spreading, roughened on the sharp angles, up to 1 m tall, up to 3 mm broad above the lowest leaf, usually slightly surpassing the leaves, brown or dark brown at base, producing leaves in the lower fourth; blades 3–4 per culm, up to 30 cm long,

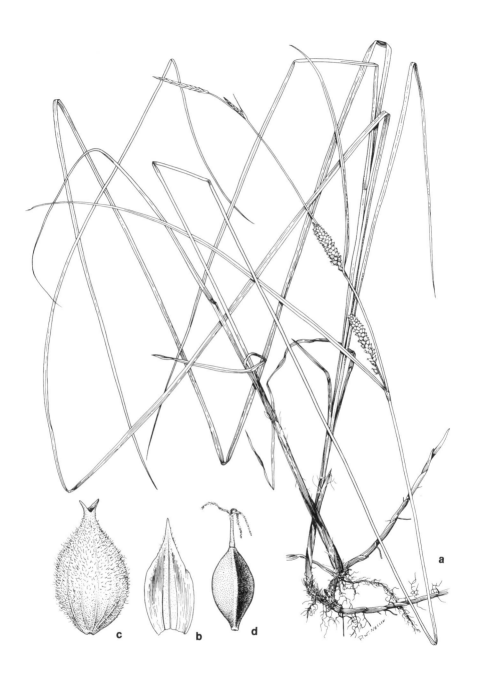

55. *Carex pellita*.

a. Habit.
b. Pistillate scale.
c. Perigynium.
d. Achene.

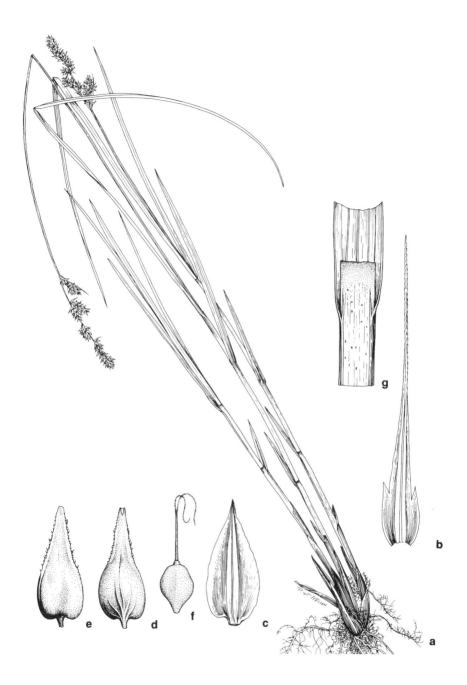

56. *Carex prairea.*
a. Habit.
b. Lowest bract.

c. Pistillate scale.
d. Perigynium, dorsal view.
e. Perigynium, ventral view.

f. Achene.
g. Sheath.

2.0–3.1 mm wide, flat but with revolute margins, ascending or spreading, papillate above, the margins and sometimes the tip harshly scabrous, light green; sheaths tight, membranous and hyaline, papillate between the pale cartilaginous veins, sometimes usually minutely copper-dotted, convex and strongly copper-tinged at the mouth, prolonged 2–3 mm beyond base of blade; inflorescence up to 10 cm long, up to 2 cm thick, continuous near the tip, usually interrupted below, compound at the lower nodes; bracts setaceous or scalelike, usually never longer than the spikes; spikes several, crowded, 3–6 mm long, 2.5–4.0 mm broad, the staminate flowers above the pistillate; pistillate scales ovate-lanceolate, up to 3.5 mm long, mostly concealing the perigynia, acute to cuspidate, reddish brown except for the 1- to 3-nerved hyaline center; staminate scales apical, inconspicuous, very narrow; perigynia narrowly ovoid, biconvex, appressed to strongly ascending, 2.5–4.0 mm long, 1.2–1.3 mm broad, strongly nerved on the lower side, dark stramineous to brown, dull, stipitate, tapering to a beak, the beak up to 1 mm long, bidentate, rough along the margins, green or whitish but usually reddish tipped; achenes lenticular, up to 2 mm long, apiculate, stipitate, jointed to a short style; stigmas 2, short, reddish brown. May–July.

Marshes, wet meadows, swamps.

IA, IL, IN, MO (FACW+), NE (OBL), OH (FACW).

Prairie sedge.

This species is distinguished from the similar *C. diandra* by its dull brown achenes and the copper-tinged mouth of the sheaths.

55. **Carex prasina** Wahl. Kongl. Vet. Acad. Handl. II, 24:161. 1803. Fig. 57.
Carex miliacea Muhl. ex Willd. Sp. Pl. 4:290. 1805.

Plants perennial, cespitose, with short rhizomes; culms slender, sharply triangular, up to 60 cm tall, scabrous, glabrous, the base often purplish; sterile shoots often present; leaves up to 5 per culm, 3–6 mm wide, flat, pale green, glabrous, scabrous on the margins, usually shorter than the inflorescence; sheaths sometimes pink or purplish; terminal spike entirely staminate or with pistillate flowers above and staminate flowers below, up to 4 cm long, up to 4 mm thick, more or less erect, on short peduncles; staminate scales lanceolate, acute to awned; lateral spikes 2–4, pistillate, up to 6 cm long, up to 5 mm thick, narrowly cylindric, ascending or pendulous, on peduncles up to 4 cm long; pistillate scales ovate, cuspidate to awned, whitish with a green center, shorter than to as long as the perigynia; perigynia up to 50 per spike, ovoid, 2.5–4.0 mm long, triangular, pale, glabrous, finely nerved, stipitate, with a green, sometimes bent beak up to 2 mm long; achenes trigonous, 1.3–1.6 mm long, pale brown, apiculate, stipitate; stigmas 3. May–June.

Wet depressions in woods, occasionally in shallow water.

IA, IL, IN, KY, MO, OH (OBL).

Graceful forest sedge.

This is the only *Carex* that has arching, pendulous, narrowly cylindrical pistillate spikes and long, curved beaks of the perigynia.

57. *Carex prasina.*

a. Habit.
b. Pistillate scale.

c. Perigynium.
d. Achene.

58. *Carex projecta.*
a. Habit.

b. Pistillate scale.
c. Perigynium, dorsal view.

d. Perigynium, ventral view.
e. Achene.

56. Carex projecta Mack. Bull. Torrey Club 35:264. 1908. Fig. 58.
Carex tribuloides Wahl. var. *moniliformis* Britt. in Britt. & Brown, Ill. Fl. 1:336. 1896.

Plants perennial, cespitose, from short, black, fibrillose rhizomes; culms to 1 m tall, sharply triangular, the angles scabrous beneath the inflorescence, rather stiff, usually shorter than the leaves, usually with several sterile culms; leaves 4–6 per fertile culm, ascending, 3–7 mm wide, flat, lax, light green, the margins and midvein scabrous in the upper half; sheaths green, the lower ones becoming stramineous or brown, the ventral band typically hyaline only near the summit; spikes 5–15 per culm, 4–8 mm long, green or stramineous, obtuse, gynecandrous, clavate at the base, all except the very uppermost separated from each other in a moniliform inflorescence up to 7 cm long; lowest bract setaceous; pistillate scales lanceolate, obtuse to subacute, reaching the base of the beak of the perigynium, stramineous with hyaline margins and a green, 3-nerved center; perigynia up to 30 in a spike, lanceolate to linear-lanceolate, plano-convex, 2.5–5.0 mm long, 1.3–1.7 mm wide, widest at or a little below the middle, membranous, stipitate, nerved on both faces, the wing diminishing abruptly below the middle, stramineous to greenish, the beak 1–2 mm long, serrulate, bidentate; achenes lenticular, 1.5–1.7 mm long, 0.5–0.7 mm wide, apiculate, stipitate, weakly continuous with the deciduous style; stigmas 2, short, red-brown. May–June.

Swampy woods, moist woods, riparian terraces.

IL, IN, MO (FACW+), KY, OH (FACW).

Moniliform sedge.

This species shows a strong similarity to *C. tribuloides* in characteristics of the perigynia and achenes but differs markedly in its separated, moniliform spikes.

57. Carex pseudocyperus L. Sp. Pl. 2:978. 1753. Fig. 59.

Plants perennial, cespitose, from short, stout rhizomes; culms rather stout, to 1 m tall, 3-angled, scabrous; leaves up to 15 mm wide, very scabrous on the margins, septate-nodulose; sheaths hyaline dorsally, yellowish ventrally, septate-nodulose; ligules as wide as long; lower 1–2 bracts leaflike, exceeding the inflorescence; terminal spike staminate, to 7 cm long, to 8 mm thick, on a short, erect peduncle; staminate scales lanceolate, rough-awned; lateral spikes 3–5, pistillate, cylindric, to 7.5 cm long, to 1.2 cm thick, with densely crowded perigynia, on slender, spreading or pendulous peduncles up to 3.5 cm long; pistillate scales linear to linear-lanceolate, rough-awned, up to 5 mm long; perigynia up to 100 per spike, lanceoloid to lance-ovoid, reflexed, with many conspicuous nerves, little if at all inflated, tapering to a short beak; beak 1.0–1.5 mm long,

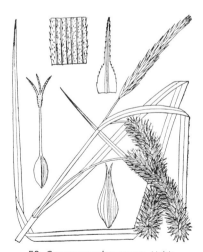

59. *Carex pseudocyperus*. Habit, section of stem, pistillate scale, achene, and perigynium.

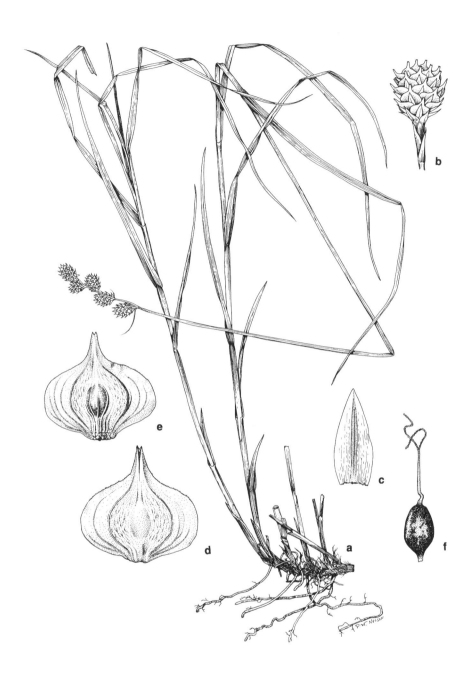

60. *Carex reniformis.*
a. Habit.
b. Spike.
c. Pistillate scale.
d. Perigynium, dorsal view.
e. Perigynium, ventral view.
f. Achene.

shorter than the body, smooth, bidentate, the teeth erect and straight, 0.5–1.0 mm long; stigmas 3; achenes trigonous, obovoid, 1.5–1.7 mm long, about 0.7 mm wide, brown. June–August.

Swamps, marshes, around ponds.

IN, OH (OBL).

Cyperus-like sedge.

This is one of the porcupine sedges that have very prickly looking pistillate spikes. It differs from *C. hystericina*, *C. lurida*, and *C. baileyi* by its non-inflated perigynia. From the very similar appearing *C. comosa*, which also has non-inflated perigynia, the perigynia are only 3–5 mm long and the teeth of the beak are straight rather than spreading.

58. **Carex reniformis** (Bailey) Small, Fl. S. E. U. S. 220. 1903. Fig. 60.
Carex straminea Willd. var. *reniformis* Bailey, Mem. Torrey Club 1:73. 1889.

Plants perennial, cespitose, from short, black, fibrillose rootstocks; culms to 1 m tall, triangular, scabrous on the angles beneath the inflorescence, light brown near the base, with old leaf bases often persisting; sterile shoots usually not present; leaves 4–5 per culm, ascending, 2.0–2.5 mm wide, flat, firm, light green, scabrous along the margins; sheaths tight, narrowly hyaline, green, prolonged at the summit; spikes 3–6 per culm, 6–10 mm long, ellipsoid to obovoid, not clavate at the base, silvery brown to green, gynecandrous, approximate but separated in a flexuous inflorescence up to 4.5 cm long, or the lower spikes distant; lowest bract setaceous, scabrous, the upper bracts scalelike; pistillate scales lanceolate, obtuse to acute, usually reaching above the base of the beak of the perigynium, the center green and 3-nerved, the margins hyaline; perigynia numerous per spike, 4–5 mm long, 3.5–5.0 mm wide, orbicular to reniform, widest at or below the middle, subcoriaceous, appressed-ascending, flat, stramineous, finely nerved on the outer face, usually nerveless on the inner face, winged nearly to base, substipitate, the beak 1.0–1.5 mm long, serrulate, bidentate, green; achenes lenticular, 1.9–2.2 mm long, about 1.5 mm wide, apiculate, stipitate; stigmas 2, reddish brown. June–September.

Wet depressions in woods.

IL (OBL).

Round-fruited sedge.

The very broad perigynia, which are often wider than they are high and are bordered by a very broad wing, are distinctive.

59. **Carex rostrata** Stokes, Bot. Arr. Brit. Pl., ed. 2, 2:1059. 1787. Fig. 61.

Plants perennial, cespitose, from long, stout rhizomes; culms up to 1.2 m tall, 3-angled, stout, usually not scabrous, pale brown or sometimes purplish at the base, with last year's leaves persistent; leaves 2–8 mm wide, flat or often canaliculate, firm, septate-nodulose, smooth or scabrous along the margins, pale green or glaucous, as long as or longer than the culms; upper sheath septate-nodulose, except for the hyaline ventral band; ligules wider than long; staminate spikes 2–4, cylindric, to 4 cm long, to 3 mm thick, pedunculate; staminate scales acute, yellow-brown with hyaline margins; pistillate spikes 2–5, cylindric to ovoid, to 7 cm long, to 1 cm

thick, sessile or the lowest one pedunculate; lower bracts leaflike, longer than the
inflorescence; pistillate scales oblong to
ovate, obtuse to acute at the tip, usually
purple-brown with hyaline margins, as
long as or longer than the perigynia;
perigynia up to 150 per spike, ovoid,
3–6 mm long, yellow-green to brown,
inflated, strongly nerved, shiny, contracted to a serrulate, bidentate beak
1–2 mm long; stigmas 3; achenes trigonous, 1.2–2.0 mm long, yellow-brown,
substipitate, continuous with the twisted
or bent style. July–September.

Bogs, sloughs, around ponds and lakes,
swamps.

IA (OBL).

Beaked sedge.

61. *Carex rostrata*. Habit pistillate scale, achene, and perigynium.

This stout species differs from *C. retrorsa* by having spreading to ascending
perigynia. It differs from *C. tuckermanii, C. oligosperma,* and *C. vesicaria* by having
the base of the culms more or less spongy and by having septate-nodulose leaves.
It differs from *C. utriculata* by its obtuse to acute pistillate scales and its somewhat
smaller perigynia.

60. **Carex sartwellii** Dewey, Am. Journ. Sci. 43:90. 1842. Fig. 62.

Plants perennial, cespitose, from fibrous roots, with thick, scaly, horizontal,
black rhizomes; culms sharply triangular, scabrous, subcanaliculate, up to 1.2 m
tall, much longer than the leaves; leaves 3–5, up to 23 cm long, 2.3–5.0 mm wide,
arising from the lower half of the culm, ascending to spreading, thinly papillose
above, the veins often appearing impressed on the lower surface, with the margins
scabrous; sheaths open, truncate to concave at the mouth, the upper green-nerved
ventrally, the lower brown to nigrescent, bladeless, with some old sheaths persisting; inflorescence narrowly cylindric, continuous or sometimes interrupted near
the base, 2.5–6.5 cm long, 0.5–1.2 cm broad, stramineous, the staminate flowers
above the pistillate; bracts lanceolate, the lowest scabrous-awned; spikes 10–25, the
upper often entirely staminate and crowded, the lower mostly pistillate and often
interrupted; pistillate scales ovate, 2.5–3.6 mm long, 1.20–1.75 mm broad, obtuse
to acute to acuminate, often erose, reddish brown, the margins hyaline; staminate
scales narrowly lanceolate, 2.5–4.0 mm long, 0.5–1.2 mm broad, acuminate, mostly
hyaline throughout; perigynia 3–25 per spike, 2.0–4.6 mm long, 1.2–2.0 mm broad,
plano-convex, ovoid-lanceoloid, substipitate, distinctly nerved dorsally and ventrally,
narrowly winged to the base, serrate in the upper third, light brown, with the
bidentate beak 0.75 mm long, serrulate, hyaline-tipped; achenes 1.30–1.75 mm
long, 0.9–1.0 mm broad, plano-convex, reddish brown, puncticulate, the margins
single-nerved, jointed to a short style; stigmas 2. April–June.

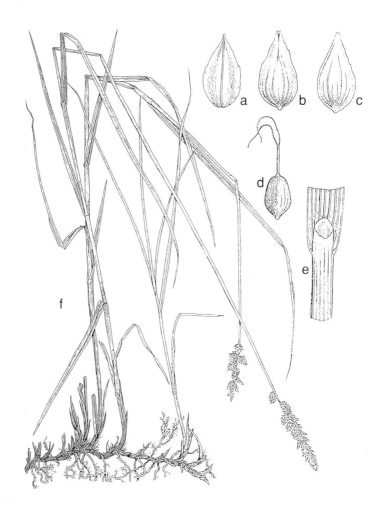

62. *Carex sartwellii.*
a. Habit.
b. Pistillate scale.

c. Perigynium, dorsal view.
d. Perigynium, ventral view.
e. Achene.

f. Sheath with ligule.

Wet prairies, calcareous wet meadows, river bottoms, marshes, dunes, peaty swamps, open cold bogs.

IA, IL, IN, MO (FACW+), NE, OH (OBL).

Sartwell's sedge.

The species differs by its solitary culms, extensive black rhizomes, and narrowly winged perigynia that are 2.0–4.6 mm long.

61. **Carex scabrata** Schwein. Ann. Lyc. N.Y. 1 (1):69. 1824. Fig. 63.

Plants perennial from long-creeping stolons, usually forming dense colonies; culms to 90 cm tall, sharply 3-angled, smooth; sterile shoots often present, with overlapping sheaths; leaves mostly on lower third of culm, up to 15 mm wide, spreading and arching, flat, scabrous along the margins, about as long as the culms; last year's sheaths dark brown, usually persistent; current year's sheaths concave at the mouth, pale on the ventral side; ligules longer than wide; lowest 1–2 bracts leaflike, exceeding the inflorescence; terminal spike staminate, to 40 cm long, to 6 mm thick, erect, short-pedunculate; staminate scales narrowly elliptic to lanceolate, pale brown or whitish with a green midnerve, up to 6 mm long; lateral spikes 3–8, pistillate, barely overlapping, to 50 mm long, to 8 mm thick, the upper sessile or nearly so, the lower on short, ascending to erect, scabrous peduncles up to 40 mm long, with up to 40 perigynia; pistillate scales lanceolate, acute at the tip,

63. *Carex scabrata.* Habit pistillate scale, and perigynium.

2.0–5.5 mm long, light brown to reddish brown with a green midnerve; perigynia 3.2–4.5 mm long, ellipsoid to obovoid, minutely hispidulous, green, few-nerved but with two prominent longitudinal ribs, stipitate, tapering to a beak, the beak 1–2 mm long, bidentate, curved at the tip; stigmas 3; achenes trigonous, obovoid, 1.2–1.7 mm long, with concave sides, brown. May–July.

Wet woods, wet meadows, along streams, rarely in shallow water.

IN, KY, MO, OH (OBL).

Scabrous sedge.

This species is readily distinguished from other wetland species of *Carex* by its short-hispidulous perigynia with a distinct, curved beak.

62. **Carex scoparia** Schk. in Willd. Sp. Pl. 4:230. 1805. Fig. 64.
Carex scoparia Schk. var. *moniliformis* Tuckerm. Enum. Meth. 8:17. 1843.
Carex scoparia Schk. var. *condensa* Fern. Proc. Am. Acad. 37:468. 1902.

Plants perennial, densely cespitose, from short, brown, fibrillose rhizomes; culms to 1 m tall, sharply triangular, the angles very scabrous above, stiffly erect, usually longer than the leaves, with sterile leafy culms uncommon; leaves up to 7 per fertile culm, ascending to spreading, 1.5–4.0 mm wide, flat, the margins and veins scabrous throughout; sheaths tight, green-nerved throughout or white-hyaline ventrally, the mouth concave, slightly thickened and sometimes darkened; spikes 3–10 per culm, 5–15 mm long, light green to stramineous, ascending, acute, gynecandrous, clavate at the base, crowded, or the lowest spike remote, or rarely all the spikes separated, forming an inflorescence 1.5–5.0 cm long, 1.0–1.5 (–2.5) cm wide; bracts setaceous; pistillate scales lanceolate, acuminate, two-thirds to three-fourths as long as and much narrower than the perigynia, tan-hyaline with the center brown to green and 3-nerved, with a central lighter nerve; perigynia many per spike, lanceolate, acuminate, plano-convex, appressed, 3.8–5.5 mm long, 1.2–2.0 mm wide, broadest

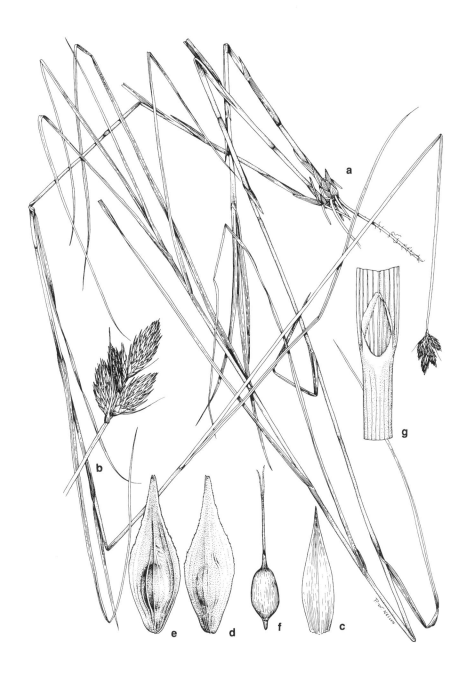

64. *Carex scoparia.*
a. Habit.
b. Inflorescence.

c. Pistillate scale.
d. Perigynium, dorsal view.
e. Perigynium, ventral view.

f. Achene.
g. Sheath.

at the middle, membranous, substipitate, distinctly nerved dorsally, obscurely nerved to nerveless ventrally, the wing not narrowed below the middle, stramineous, the beak about 3 mm long, serrulate, bidentate, the teeth appressed; achenes lenticular, 1.25–1.50 mm long, 0.75–1.00 mm wide, apiculate, stipitate, brown, weakly continuous with the deciduous, jointed style; stigmas 2, short, tan to reddish, flexuous. May–July.

Wet open woods, wet prairies, wet meadows, seeps, calcareous fens, swamps. IA, IL, IN, KS, KY, MO, NE, OH (FACW).

Wing-fruited sedge.

The field characters of this species are its very slender perigynia, narrow leaves, and pointed spikes. Plants with very crowded spikes may be called var. *condensa*. Plants with moniliform spikes may be referred to as var. *moniliformis*.

63. **Carex seorsa** Howe, Fl. Renss. Co. 39. 1894. Fig. 65.
Carex rosaeoides Howe in Gordon & Howe,
Fl. Renss. Co. 39. 1894.

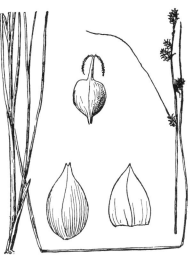

Plants perennial, loosely cespitose, from short rhizomes; culms 3-angled, smooth or scabrous only beneath the inflorescence, soft, weak, to 75 cm tall; leaves 3–5 per culm, 2–4 mm wide, flat, green, smooth or slightly scabrous along the margins, shorter than the culms; sheaths tight, smooth, the inner band hyaline, concave at the summit, the lower ones usually brownish; inflorescence up to 7 cm long, with 2–7 sessile, separated spikes; terminal spike clavate, up to 20 mm long, up to 6 mm thick, gynecandrous, the pistillate part 5- to 20-flowered; lateral spikes pistillate or occasionally with a few staminate flowers at the base, sessile, up to 10 mm long, up to 5 mm thick; pistillate scales ovate, 1.0–2.5 mm

65. *Carex seorsa.* Habit, achene, pistillate scale, and perigynium.

long, stramineous, with a green or brown center, obtuse at the tip; perigynia 2–3 mm long, 1.0–1.8 mm wide, ellipsoid to narrowly ovoid, plano-convex, conspicuously nerved, more or less spongy-thickened at the substipitate base, green, tapering to a beak, at least the lowest perigynia spreading to reflexed; beak of the perigynium 0.3–0.7 mm long, not serrulate, jointed to the deciduous style; stigmas 2; achenes plano-convex. April–July.

Swamps, wet woods.

IN (FACW+), OH (FACW).

Soft-stem sedge.

This species differs from all other sedges with star-shaped spikes, including *C. interior, C. atlantica*, and *C. echinata*, by the smooth beak of the perigynium.

64. **Carex squarrosa** L. Sp. Pl. 2:973. 1753. Fig. 66.

Plants perennial, cespitose, from short rhizomes; culms to 90 cm tall, triangular, nearly smooth, usually dark brown at the base; sterile shoots usually present; leaves up to 6 mm wide, dark green, firm, slightly scabrous along the margins, longer than the culms; sheaths tight, pale brown, the lowermost fibrous and red-tinged, the ligule longer than wide; spike usually solitary, sometimes 2–4, up to 3 cm long, up to 2.2 cm thick, the terminal one subglobose to thick-cylindric, pistillate except for several staminate flowers at the base; staminate scales acute to awned, with a green center and hyaline margins; lateral spikes, if present, 2–4, entirely pistillate, subglobose to thick-cylindric; pistillate scales acute to acuminate to cuspidate, usually reaching the base of the beak of the perigynium; perigynia up to 100 or more per spike, obconic, rounded across the top, the lowermost reflexed, 4–5 mm long, up to 3 mm broad, pale green to pale brown, many-nerved, abruptly contracted into a bidentate beak to 3.5 mm long; achenes trigonous, 2.5–3.2 mm long, black, substipitate; stigmas 3. April–September.

Swamps, marshes, fens, around ponds, sinkhole ponds.

IA, IL, IN, MO (OBL), KS, KY, OH (FACW).

Squarrose sedge.

This species is similar in appearance to *C. typhina*, but the pistillate scales of *C. squarrosa* are acuminate to cuspidate, while those of *C. typhina* are obtuse. Some of the lowest perigynia in *C. squarrosa* are reflexed.

65. **Carex sterilis** Willd. Sp. Pl. 4:208. 1805. Fig. 67.
Carex muricata L. var. *sterilis* (Willd.) Gl. Phytologia 4:22. 1952.

Plants perennial, densely cespitose, from stout rhizomes; culms wiry, sharply triangular, scabrous, to 75 cm tall; leaves 3–5, 1.5–2.5 mm wide, flat, firm, green, scabrous along the margins, shorter than the culms; sheaths tight, smooth, green, the inner band hyaline, usually minutely papillate, concave at the apex, at least the lowermost brown; inflorescence up to 4 cm long, densely crowded above but with the lower spikes often separated, with 3–8 sessile spikes, usually dioecious, or with pistillate plants often bearing a very few staminate flowers, or staminate plants with a few scattered perigynia, or sometimes with all staminate spikes and all pistillate spikes on the same plant; terminal spike 3.5–12.0 mm long, usually unisexual; lateral spikes up to 12 mm long, subtended by scalelike bracts; pistillate scales ovate, 2–3 mm long, castaneous with a green center and hyaline margins, acute at the tip, reaching to the base of the perigynium or longer; perigynia 2.0–3.5 mm long, 1.0–2.2 mm wide, lanceoloid to deltoid, plano-convex, the ventral surface with up to 10 nerves, or nerveless, spongy-thickened at the base, castaneous to nearly black, sessile, tapering to a beak, the perigynia radiating in all directions; beak up to 1.5 mm long, serrulate, bidentate at the apex; achenes biconvex, 1.0–1.7 mm long, sessile, jointed to the deciduous style; stigmas 2. April–May.

Wet meadows, fens, marly seeps.

IL, KY, OH (OBL).

Sterile sedge.

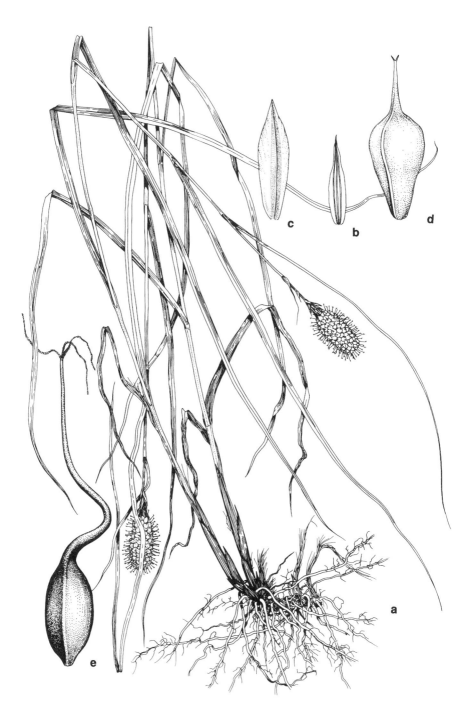

66. *Carex squarrosa.*
a. Habit.

b. Staminate scale.
c. Pistillate scale.

d. Perigynium.
e. Achene.

67. *Carex sterilis.*
a. Habit.
b. Inflorescence.
c. Pistillate scale.
d. Perigynium, dorsal view.
e. Perigynium, ventral view.
f. Achene.
g. Sheath with ligule.

This is one of the species that has the perigynia radiating in all directions so that the spikes resemble a star. *Carex sterilis* differs from all the other star sedges by its unisexual terminal spike.

66. **Carex stipata** Muhl. in Willd. Sp. Pl. 4:233. 1805. Fig. 68.
Carex stipata Muhl. var. *maxima* Chapm. ex Boott, Ill. Carex 122, pl. 391. 1862.

Plants perennial, cespitose, with fibrous roots and short, stout, dark rhizomes; culms up to 1.2 cm long, strongly 3-angled, spongy, easily compressed, rough to the touch, light brown at base, usually overtopped by the leaves; leaves 3–6, up to 1 m long, 4–15 mm wide, soft, flaccid, often yellowish green, rough along the margins and on the midvein beneath; sheaths open, green, prolonged at the summit, septate-nodulose; ligule longer than wide; inflorescence elongated, compound, usually continuous, up to 15 cm long, up to 4 cm broad; bracts setaceous, up to 5 cm long; spikes 15–25, the staminate flowers above the pistillate; pistillate scales lanceolate to ovate, acuminate to cuspidate to short-awned, hyaline to yellow-brown, with a green midnerve, much shorter than the perigynia; staminate scales narrowly lanceolate, pale brown; perigynia 4–10 per spike, lanceoloid, 3.5–6.0 mm long, 1.5–2.0 mm broad, plano-convex, rounded and spongy thickened at base, stipitate, yellowish or pale brown, conspicuously few-nerved, tapering gradually into a 2-toothed, serrulate green beak 2.0–3.5 mm long; achenes 1.5–2.0 mm long, lenticular, ovoid to orbicular, apiculate, substipitate, jointed to the style, the style swollen at base; stigmas 2, reddish brown. May–July.

Marshes, around ponds and lakes, fens, ditches, along rivers and streams, swamps. IA, IL, IN, KS, KY, MO, NE, OH (OBL).

Spongy sedge.

Carex stipata is easily identified by its long, slender, spongy-based perigynia and its spongy, strongly 3-angled culms. Specimens in which most parts of the plant are larger may be called var. *maxima*. This species differs from *C. laevivaginata* by its brownish or yellowish perigynia, its pistillate scales shorter than the perigynia, and its longer and thicker inflorescences.

67. **Carex straminea** Willd. in Schk. Riedgr. 49. 1801. Fig. 69.
Carex tenera Dewey var. *richii* Fern. Proc. Am. Acad. 37:475. 1902.
Carex richii (Fern.) Mack. Bull. Torrey Club 49:362. 1923.

Plants perennial, cespitose, from short, black, fibrillose rootstocks; culms to 1 m tall, sharply triangular, the angles scabrous beneath the inflorescence, rather stiff, usually about as long as the leaves, usually with few or no sterile culms; leaves 3–5 per culm, ascending, 2.0–2.5 mm wide, flat, rather lax, green, the margins sca-brous; sheaths tight, green-nerved throughout except for an abbreviated ventral band near the summit, the lowermost stramineous; spikes 3–8 per culm, 6–10 mm long, green or stramineous, obtuse, gynecandrous, clavate at base, the uppermost sometimes crowded, otherwise the spikes separated in an elongated inflorescence up to 8 cm long; lowest bract setaceous or scalelike; pistillate scales lanceolate, acuminate to aristate, the tip of the uppermost surpassing the base of the beak of the perigynium, tan with hyaline margins and a green center; perigynia up to 30 in

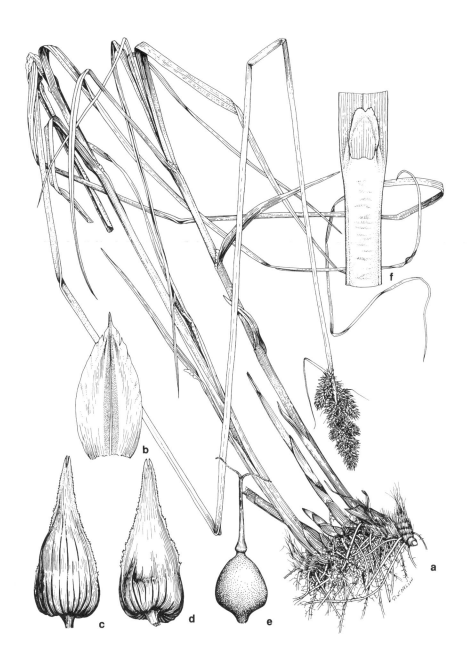

68. *Carex stipata.*
a. Habit.
b. Pistillate scale.

c. Perigynium, dorsal view.
d. Perigynium, ventral view.
e. Achene.

f. Sheath with ligule.

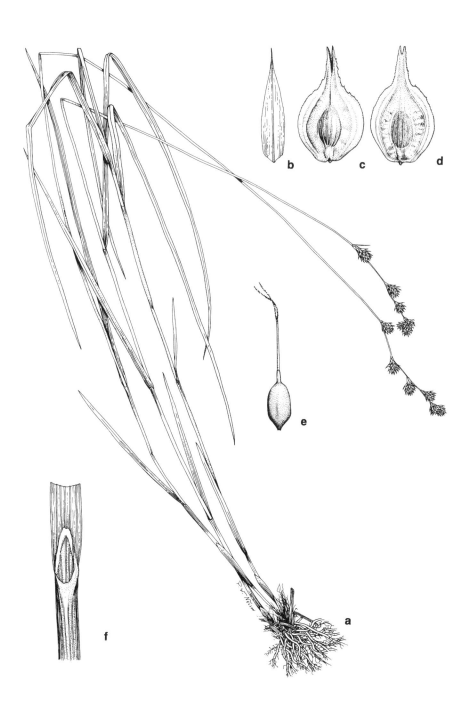

69. *Carex straminea.*
a. Habit.
b. Pistillate scale.

c. Perigynium, dorsal view.
d. Perigynium, ventral view.
e. Achene.

f. Sheath with ligule.

a spike, 2.5–3.5 mm long, 1.5–2.7 mm wide, ovate, widest at the middle, appressed-ascending, membranous, plano-convex, strongly nerved on both faces, winged to the base, stramineous to greenish, the beak 1.5–2.0 mm long, serrulate, bidentate; achenes lenticular, 1.5–1.7 mm long, 0.75–1.00 mm wide, apiculate, stipitate, weakly jointed with the deciduous style; stigmas 2, short, reddish. April–June.

Ditches, in sinkhole ponds.

IA, IL, IN, KY, MO, OH (OBL).

Straw-colored sedge.

This species is distinguished by its awn-tipped pistillate scales, its ovate perigynia, and its moniliform inflorescence.

68. **Carex stricta** Lam. Encycl. 3:387. 1792. Fig. 70.
Carex strictior Dewey in Wood, Class-Book 582. 1845.
Carex stricta Lam. var. *strictior* (Dewey) Carey in Gray, Man. Bot. 548. 1848.

Plants perennial, cespitose, producing scaly, brown stolons; sterile shoots common; culms to 1.5 m tall, firm, sharply triangular, scabrous on the angles, red-brown at the base, often with last year's leaves persistent; leaves up to 6 mm wide, stiff, keeled, often revolute, usually scabrous along the margins, often glaucous when young, usually shorter than the culms; lowest sheaths bladeless, becoming fibrous, brown or nearly black to reddish or green, smooth or scabrous, often glaucous when young, the ligule longer than the width of the blade, forming a sharp V; upper 2–3 spikes staminate, up to 6.5 cm long, erect, on short peduncles, densely flowered; staminate scales brown, often tinged with purple; lower 3–4 spikes pistillate but with a few staminate flowers at the tip, up to 10 cm long, erect, subsessile or on short peduncles up to 1.5 cm long; pistillate scales ovate, obtuse to acute, 1.5–3.5 mm long, red-brown or hyaline and brown-dotted, shorter than the perigynia; perigynia numerous, more or less flat, 1.7–3.4 mm long, pale brown, ovate to elliptic, broadest at or below the middle, appressed-ascending, 2-ribbed, otherwise nerveless, often granular-papillate, with a minute beak 0.1–0.2 mm long, substipitate; achenes lenticular, 1.0–1.8 mm long, dull brown, short-apiculate, substipitate; stigmas 2. May–July.

Marshes, fens, along streams, swamps.

IA, IL, IN, KS, KY, MO, NE, OH (OBL).

Tussock sedge.

This species often forms dense, huge clumps. It differs from the similar *C. emoryi* in its V-shaped ligule and from *C. haydenii* in its shorter pistillate scales and its purple-tinged staminate scales.

69. **Carex suberecta** (Olney) Britt. Man. Fl. N. States, ed. 2, 1057. 1905. Fig. 71.
Carex tenera Dewey var. *suberecta* Olney, Caric. Bor. Am. 3. 1871.

Plants perennial, densely cespitose, from short, black, fibrillose rootstocks; culms to 1 m tall, sharply triangular, the angles scabrous, at least beneath the inflorescence, light brown at the base, with the remains of previous year's leaves usually present; leaves 3–5 per culm, ascending, 1.5–3.3 mm wide, firm, light green, scabrous along the margins; sheaths more or less tight, green-nerved nearly throughout,

70. *Carex stricta*.

a. Habit.
b. Pistillate scale.

c. Perigynium.
d. Achene.

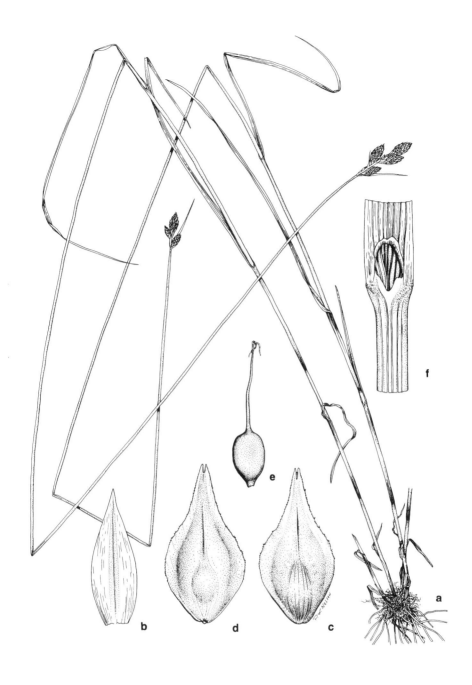

71. *Carex suberecta.*
a. Habit.
b. Pistillate scale.

c. Perigynium, dorsal view.
d. Perigynium, ventral view.
e. Achene.

f. Sheath with ligule.

prolonged at the summit, the lowermost stramineous to brown; spikes 2–5 per culm, 7–12 mm long, tapering to each end, yellow-brown, approximate but distinct in an inflorescence 1.5–3.0 cm long, gynecandrous; lowest bract cuspidate, upper bracts scalelike; pistillate scales lanceolate, long-acuminate to subaristate, usually reaching the base of the beak of the perigynium, the center green and 3-nerved, the margins hyaline; perigynia numerous per spike, 4–5 mm long, 2.0–2.8 mm wide, rhombic, widest above the middle, appressed, plano-convex, yellow-brown, faintly nerved on the outer face, nerveless or with 1–5 faint nerves on the inner face, the wings diminishing toward the base, substipitate, the beak 0.5–1.5 mm long, serrulate, bidentate, usually greenish; achenes lenticular, 1.5–1.8 mm long, about 1 mm wide, apiculate, substipitate; stigmas 2, short, reddish brown. May–July.

Fens, seeps.

IL, MO, OH (OBL).

Upright sedge.

This species is recognized by its rather crowded spikes and its moderately large perigynia that are widest above the middle and more or less cuneate at the base.

70. **Carex sychnocephala** Carey, Am. Journ. Sci. & Arts, ser. 2, 4 (10):24. 1847. Fig. 72.

Plants perennial, cespitose, from short, brown, fibrillose rhizomes; culms rather stout, up to 50 cm tall, 3-angled with blunt angles, smooth, erect, longer than the leaves; leaves up to 5 per culm, ascending to spreading, 2–4 mm wide, flat, soft, usually smooth; sheaths tight, usually several-nerved and hyaline ventrally, the mouth concave; lower bracts leaflike, erect, up to 20 cm long, forming an involucre; spikes 4–10 per culm, ellipsoid to oblongoid, to 15 mm long, gynecandrous, many-flowered, green or stramineous, crowded and forming a dense head; pistillate scales linear to narrowly lanceolate, long-acuminate, several-nerved, shorter and narrower than the perigynia; perigynia subulate to narrowly lanceolate, long-acumi-

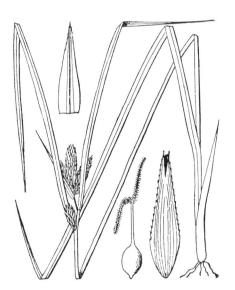

72. *Carex sychnocephala.* Habit, pistillate scale, achene, and perigynium.

nate at the tip, plano-convex, appressed, 4.5–5.0 mm long, 0.7–0.9 mm wide, broadest below the middle, substipitate, distinctly few-nerved, the wing not reaching the base, green to stramineous, the serrulate beak 2–3 times longer than the body, bidentate; achenes lenticular, 2–3 mm long, 0.3–0.5 mm wide, apiculate, stipitate, brownish, continuous with the deciduous, jointed style; stigmas 2, short, pale brown. July–August.

Wet meadows, occasionally in shallow water.

IA, MO (FACW+).

Long-beaked sedge.

This is the only member of the Ovales group of *Carex* that has a dense, headlike cluster of spikes subtended by long, erect, leafy bracts that form an involucre.

71. **Carex tenera** Dewey, Am. Journ. Sci. 8:97. 1824. Fig. 73.

Carex stramineae Willd. var. *echinodes* Fern. Proc. Am. Acad. 37:474. 1902.

Carex tenera Dewey var. *echinodes* (Fern.) Wieg. Rhodora 26:2. 1924.

Plants perennial, densely cespitose, from short, black, scaly and fibrillose rhizomes; culms to 70 cm tall, sharply triangular, scabrous beneath the inflorescence or smooth, stiff, longer than the leaves, dark brown to black at the base, with the old bases often remaining as stubble; leaves 3–5 per culm, ascending, 1.0–2.8 mm wide, flat, green, the margins scabrous toward the apex; sheaths tight, green-nerved nearly throughout or with a narrow, hyaline, ventral band, the mouth prolonged and concave, the lowermost sheath purplish brown to blackish; spikes 3–7 per culm, 6–10 mm long, green to stramineous, obtuse, gynecandrous, somewhat clavate at base, the staminate flowers conspicuous, the spikes arranged in a moniliform inflorescence to 5 cm long, the axis of the inflorescence narrowed and flexed just above the lower spike; lowest bract usually setaceous, scabrous, the upper bracts scalelike; pistillate scales ovate to ovate-lanceolate, acute to acuminate, reaching or surpassing the base of the beak of the perigynium, the center greenish and 3-nerved, the margins hyaline; perigynia many per spike, 3.0–4.4 mm long, 1.5–2.0 mm wide, ovate, widest a little below the middle, spreading or ascending, plano-convex, strongly nerved on the outer face, more finely nerved on the inner face, narrowly and evenly winged to the base, substipitate, the beak 0.8–1.2 mm long, serrulate, bidentate; achenes lenticular, 1.5–1.8 mm long, 1.0–1.3 mm wide, apiculate, stipitate, weakly jointed with the deciduous style; stigmas 2, slender, light reddish. May–June.

Floodplain woods, wet meadows, mesic prairies, swampy depressions, wet ditches.

IA, IL, IN, MO (FAC+), KS, NE (FACW), KY, OH (FAC).

Remote sedge.

The combination of remote, moniliform, gynecandrous spikes and ovate perigynia is distinct for this species. Specimens with perigynia 4.0–4.4 mm long may be known as var. *echinodes*.

72. **Carex tetanica** Schk. Riedgr. Nachtr. 68. 1806. Fig. 74.

Plants perennial, with elongated, slender rhizomes and numerous white stolons; culms slender, triangular, up to 60 cm tall, scabrous, usually purplish at the base; leaves up to 5 per culm, up to 5 mm wide, green, thin, scabrous along the margins; sheaths often reddish purple, the lowest usually blade-bearing; terminal spike staminate, up to 4 cm long, up to 3.5 mm thick, on a scabrous peduncle up to 4 cm long; staminate scales obtuse, red-purple with a hyaline margin and a green center; lateral spikes 1–3, pistillate, to 4 cm long, to 5 mm thick, at least some of them on scabrous peduncles; pistillate scales ovate, obtuse to acute, purple-brown with a

73. *Carex tenera.*
a. Habit.
b. Inflorescence.

c. Pistillate scale.
d. Perigynium, dorsal view.
e. Perigynium, ventral view.

f. Achene.

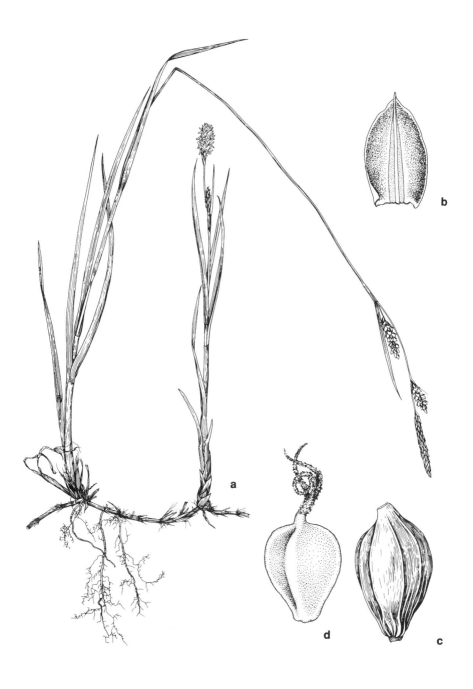

74. *Carex tetanica.*

a. Habit.
b. Pistillate scale.
c. Perigynium.
d. Achene.

hyaline margin and a green center, shorter than to about as long as the perigynia; perigynia up to 20 per spike, ovoid, 2.5–3.5 mm long, trigonous, not turgid, strongly nerved, glabrous, usually curved at the nearly beakless tip; achenes trigonous, 2.0–2.5 mm long, brown, apiculate; stigmas 3. April–June.

Wet prairies, fens, wet meadows.

IA, IL, IN, OH (FACW), NE (FACW+).

Sedge.

The tip of the perigynium is slightly curved, although the beak is miniscule. The culms are usually purplish at the base.

73. **Carex torta** Boott ex Tuckerm. Enum. Meth. 11. 1843. Fig. 75.

Plants perennial, cespitose, usually with slender rhizomes; sterile shoots common; culms to 75 cm tall, usually smooth on the angles, red-purple at the base, with last year's leaves persistent; leaves 3–5 mm wide, flat, scabrous along the margins and on the upper surface, shorter than the culms; lowest sheaths bladeless, red-tinged, the upper ones with sheaths green-nerved, with a broad, usually russet-spotted hyaline ventral band; terminal 1–2 spikes staminate, erect, up to 4.5 cm long; staminate scales red-brown with a hyaline margin and paler midvein; lateral spikes 3–6, pistillate or with a few staminate flowers at the tip, up to 9 cm long, ascending to pendulous, sessile or subsessile or on capillary peduncles; pistillate scales elliptic, obtuse, purple-black with a hyaline margin and a green midvein, barely reaching the base of the perigynium; perigynia numerous, 2.5–4.5 mm long, flat to plano-convex, ovate to obovate, deep green, 2-ribbed, otherwise nerveless, contracted into a short, twisted beak, substipitate; achenes lenticular, dull brown, 1.8–2.0 mm long, substipitate; stigmas 2. April–June.

In flowing water of streams.

IL, MO (OBL), KY, OH (FACW).

Twisted sedge.

This species is readily distinguished by its nerveless perigynium, each with a minute, twisted beak, and by its purple-black pistillate scales with hyaline margins and green midveins.

74. **Carex triangularis** Boeck. Flora 39:226. 1856. Fig. 76.

Plants perennial, cespitose, from short, black rhizomes; culms to 85 cm tall, rather slender, 3-angular, smooth or scabrous; leaves to 4.5 mm wide, scabrous along the margins, longer than the culms; lower sheaths tight, truncate to concave at the summit, cross-rugulose, usually red-dotted on the ventral side; inflorescence elongated, to 50 mm long, to 18 (–20) wide, with 15 or more crowded spikes; spikes sessile, to 7 mm long, to 6 mm thick, androgynous; scales lance-ovate, long-acuminate or often awned, 1.8–3.0 mm long; bracts setaceous, short; perigynia deltoid to ovoid to suborbicular, 3.0–3.5 mm long, 2.5–3.0 mm wide, short-beaked at the apex, broadly rounded or truncate at the base, obscurely nerved or nerveless, horizontally spreading, concealing the scales; stigmas 2; achenes lenticular, 1.5–2.0 mm long. May–June.

75. *Carex torta.* a. Habit. c. Perigynium.

b. Pistillate scale. d. Achene.

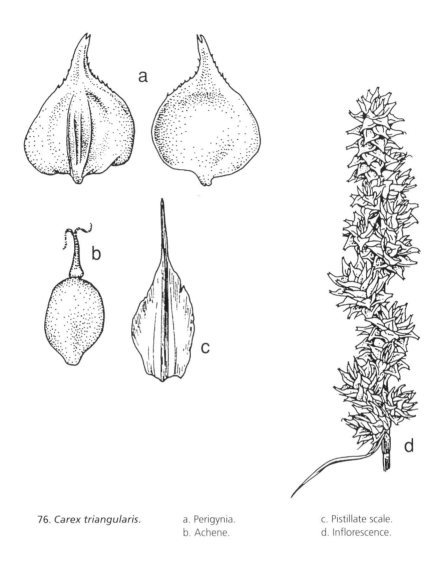

76. *Carex triangularis.*

a. Perigynia.
b. Achene.

c. Pistillate scale.
d. Inflorescence.

Swamps, bottomland forests, ditches.
KS (FACW), KY (NI), MO (OBL).
Triangular sedge.

Most plants of *C. triangularis* have some of the lower sheaths red-dotted. The wide-spreading perigynia usually completely obscure the pistillate scales.

75. **Carex tribuloides** Wahl. Sv. Vet. Akad. Handl. 24:145. 1803. Fig. 77.

Plants perennial, densely cespitose, from short, brown to black, fibrillose rhizomes; culms to 1.2 m tall, sharply triangular, the angles scabrous, stiffly erect or somewhat lax, usually about as long as the leaves, with sterile leafy culms common; leaves 5–9 per fertile culm, ascending to spreading, 3.5–7.0 mm wide, flat, the margins scabrous in the upper half; sheaths tight, the upper greenest near the summit, the lower pale or becoming stramineous or brown, the ventral portion veiny nearly throughout; spikes 4–12 per culm, 6–12 mm long, pale green to stramineous, ascending, acute to obtuse, gynecandrous, clavate at the base, silvery when immature, mostly aggregated, or with the lowest 1–3 spikes slightly separated, forming an inflorescence 1.5–6.0 cm long; bracts setaceous; pistillate scales lanceolate, acute to acuminate, two-thirds to three-fourths as long as and narrower than the perigynia, tan-hyaline with the center greenish to darker brown and 3-nerved; perigynia many per spike, lanceolate to narrowly lanceolate-ovate, plano-convex, appressed to slightly spreading, 3.25–5.00 mm long, 1.25–1.50 mm wide, widest at the middle, membranous, substipitate, discretely nerved dorsally and ventrally, the wing abruptly narrowed to absent below the middle, pale green to stramineous, the beak 1.0–1.5 mm long, serrulate, bidentate, the teeth appressed; achenes lenticular, 1.5–1.7 mm long, 0.75–1.00 mm wide, apiculate, stipitate, light to dark brown, weakly continuous with the deciduous style; stigmas 2, slender, elongate, reddish. April–August.

Wet woods, swamps, wet ditches, peaty marshes, swales, wet prairies, wet meadows, peaty fens, oxbows, shores of lakes and ponds, sometimes in standing water.

IA, IL, IN, KY, MO, OH (FACW+), KS, NE (FACW).

Narrow-fruited sedge.

This species is distinguished by its crowded spikelets, very narrow perigynia, and leaves that are at least 3 mm broad. The perigynia are always less than 2 mm wide and cuneate rather than rounded at the base. The spikes are longer than broad. The sterile culms have leaves that are often strongly 3-ranked, thereby resembling the sterile culms and leaves of *C. muskingumensis*.

76. **Carex trichocarpa** Muhl. ex Schk. Nachtr. Rieddgr. 47. 1806. Fig. 78.
Carex trichocarpa Muhl. var. *imberbis* Gray, Man. Bot., ed. 5, 597. 1867.

Plants perennial, cespitose, with slender rhizomes; culms to 1.2 m tall, triangular, scabrous, purplish at the base; sterile shoots usually present; leaves up to 8 mm wide, dull green, glabrous, septate-nodulose, scabrous along the margins; sheaths tight, green, septate, the ventral band reddish, the lowermost red and bladeless, the ligule as wide as long; upper 2–5 spikes staminate, up to 5 cm long, up to 4 mm thick, all sessile except sometimes the uppermost; staminate scales obtuse, awned, pale brown, with hyaline margins; pistillate spikes 2–4, elongate-cylindric, up to 9 cm long, up to 1.5 cm thick, the upper sessile, the lowest on a stiff, erect, scabrous peduncle; bracts leaflike; pistillate scales broadly ovate, acute to awned, reddish brown with hyaline margins, shorter than the perigynia; perigynia up to 40 per spike, ovoid, 5–10 mm long, 2.5–3.0 mm wide, ovoid, pubescent, strongly nerved, subcoriaceous, dull green or stramineous, tapering to a bidentate beak 3.0–3.5 mm long, the teeth 1.4–2.0 mm long; achenes trigonous, 2.2–2.5 mm long, yellowish, usually substipitate; stigmas 3. June–August.

Calcareous wet meadows, sloughs, seeps, marshes.

IA, IL, IN, OH (OBL).

Hairy-fruited beaked sedge.

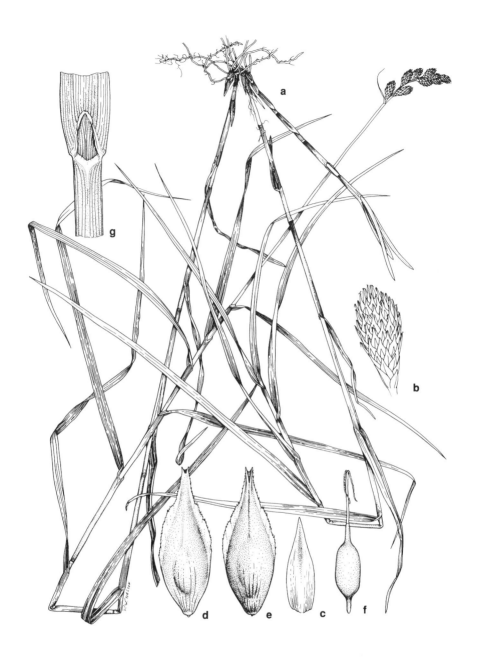

77. *Carex tribuloides.*
a. Habit.
b. Spike.
c. Pistillate scale.
d. Perigynium, dorsal view.
e. Perigynium, ventral view.
f. Achene.
g. Sheath with ligule.

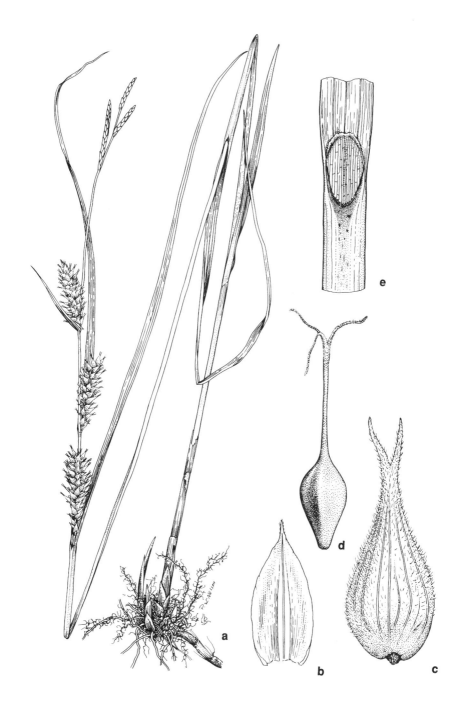

78. *Carex trichocarpa.*
a. Habit.
b. Pistillate scale.

c. Perigynium.
d. Achene.
e. Sheath with ligule.

f. Sterile shoot.

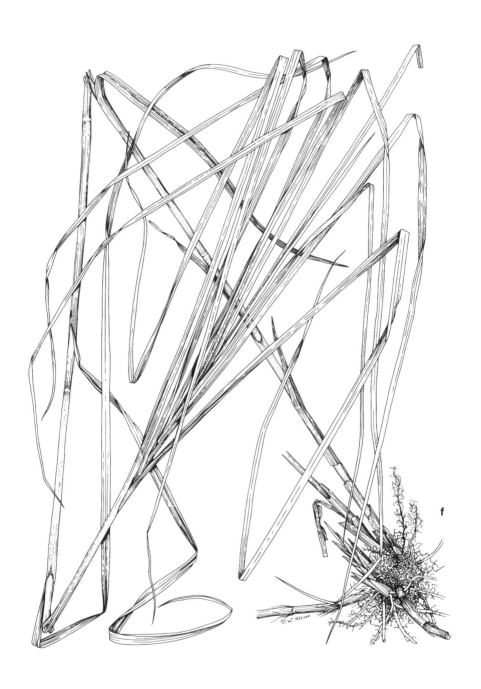

f

Carex trichocarpa is distinguished by its several staminate spikes, its strongly nerved, pubescent perigynia, and the long teeth of its perigynial beaks.

77. **Carex trisperma** Dewey, Am. Journ. Sci. 9:63. 1825. Fig. 79.

Plants perennial, loosely cespitose, with fibrous roots and slender, pale brown rhizomes; culms weakly erect to reclining, up to 70 cm tall, very slender, smooth or rough to the touch beneath the inflorescence, brown at base, with several old leaves persistent at base; leaves 3–5, up to 20 cm long, 0.75–2.00 mm wide, weak, flat or canaliculate, dark green, rough along the margins, much shorter than the culms; sheaths open, tight, concave to truncate at the summit, hyaline; ligule longer than wide; inflorescence composed of a few sparsely flowered, interrupted spikes, up to 6.5 cm long, up to 0.5 cm broad; lowest bracts filiform to setaceous, up to 10 cm long, as long as or longer than the inflorescence, the upper ones reduced to awned or aristate scales; spikes 1–4 per culm, few-flowered, the upper 2–3 often aggregated, the lowest 1–2 remote, the staminate flowers below the pistillate flowers; pistillate scales ovate to ovate-lanceolate, acute to acuminate to aristate, hyaline with 3 green nerves, shorter and barely narrower than the perigynia; staminate scales narrowly lanceolate; perigynia 1–5 per spike, oblongoid, 2.5–4.0 mm long, 1.25–2.00 mm broad, ascending, plano-convex, thick-coriaceous to spongy, substipitate, olive to light green to brown, white-puncticulate, sharp-edged along the margin, finely nerved, abruptly tapering to a very short, untoothed, green beak less than 0.50–0.75 mm long; achenes 2 mm long, lenticular, yellowish brown, shiny, apiculate, substipitate, jointed to the style; stigmas 2, reddish brown, slender. June–July.

Tamarack swamps, sphagnum bogs.

IL, IN, OH (OBL).

Three-seeded sedge.

Carex trisperma has a rather depauperate appearance because the inflorescence consists of only 1–4 spikes, each with only 1–5 perigynia. It is similar to *C. disperma*, but the staminate flowers in *C. trisperma* are below the pistillate flowers.

78. **Carex tuckermanii** Boott ex Dewey, Am. Journ. Sci. 49:48. 1845. Fig. 80.

Plants perennial, loosely cespitose, with short, stout rhizomes; culms to 1 m tall, triangular, slender, usually scabrous beneath the inflorescence, purplish red at the base; sterile shoots present; leaves 3–5 mm wide, septate-nodulose, dark green, scabrous along the margins; sheaths septate-nodulose, the lower reddish, the ventral band nerveless, the ligule as wide as long; staminate spikes 2–3, to 5 cm long, to 2.5 mm thick, stramineous, separated, sessile or short-pedunculate; staminate scales obtuse, yellow-brown to purple-brown, with hyaline margins and a green center; pistillate spikes 2–3, thick-cylindric, up to 6 cm long, up to 1.8 cm thick, sessile or on slender peduncles; bracts leaflike; pistillate scales broadly lanceolate, acuminate, often aristate, yellow-brown to purple-brown, with hyaline margins and a green center, just reaching the base of the beak of the perigynium; perigynia up to 30 per spike, 7–10 mm long, 4.5–6.5 mm wide, the body suborbicular to broadly ovate, membranous, inflated, golden brown, shiny, strongly few-nerved, tapering to a smooth, bidentate beak 2.5–3.0 mm long, the teeth 1–2 mm long; achenes trigo-

79. *Carex trisperma*.
a. Habit.
b. Spike.

c. Pistillate scale.
d. Perigynium, dorsal view.
e. Perigynium, ventral view.

f. Achene.

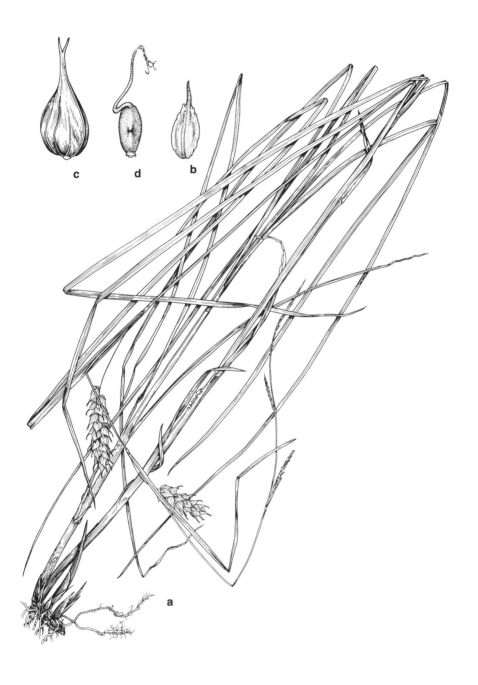

c

d

b

a

80. *Carex tuckermanii.* a. Habit.
b. Pistillate scale.

c. Perigynium.
d. Achene.

nous, 3–4 mm long, 2.0–2.5 mm wide, yellowish, deeply grooved on one side, substipitate, continuous with the persistent, twisted or bent style; stigmas 3. June–August.

Upland depressions in wet savannas.

IA, IN, IL, OH (OBL).

Tuckerman's sedge.

This species is readily distinguished by its inflated, bladderlike perigynia and its achenes that are deeply grooved on one side.

79. **Carex typhina** Michx. Fl. Bor. Am. 2:169. 1803. Fig. 81.

Plants perennial, cespitose, from short rhizomes; culms to 1 m tall, triangular, nearly smooth, brown at the base; sterile shoots usually present; leaves up to 10 mm wide, dark green, firm, somewhat scabrous along the margins, longer than the culms; sheaths tight, pale brown, the lowermost red-tinged, the ligule longer than wide; spikes 2–5, up to 4 cm long, up to 1.7 cm thick, the terminal one thick-cylindric, gynecandrous; staminate scales obtuse, reddish brown with a green center; lateral spikes entirely pistillate, thick-cylindric; pistillate scales obtuse, or the uppermost acute, shorter than to just reaching the base of the beak of the perigynium; perigynia up to 100 or more per spike, obconic, rounded to nearly truncate across the top, 4–5 mm long, up to 3 mm broad, none of them reflexed, pale green to pale brown, many-nerved, abruptly contracted into a bidentate beak to 3.5 mm long; achenes trigonous, 2.2–2.5 mm long, black, substipitate; stigmas 3. June–September.

Bottomland woods, swamps, wet meadows.

IA, IL, IN, MO (FACW+), KY, OH (OBL).

Cat-tail sedge.

This species differs from the similar *C. squarrosa* in its several more slender spikes, its usually obtuse pistillate scales, and the perigynia, none of which is reflexed.

80. **Carex utriculata** Boott in Hook. Fl. Bor. Am. 2:221. 1839. Fig. 82.
Carex rostrata Stokes var. *utriculata* (Boott) Bailey, Proc. Am. Acad. 22:67. 1886.

Plants perennial, cespitose, from long, stout rhizomes; culms up to 1.2 m tall, triangular, stout, usually not scabrous, pale brown or sometimes purplish at the base, with last year's leaves persistent; leaves 6–12 mm wide, firm, septate-nodulose, smooth or scabrous along the margins, yellow-green, usually as long as or longer than the culms; upper sheaths septate-nodulose except for the hyaline ventral band, the lower sheaths spongy, pale brown to red-tinged, the ligule wider than long; staminate spikes 2–4, to 6 cm long, to 3 mm thick, pedunculate; staminate scales acute, yellow-brown with hyaline margins; pistillate spikes 2–5, thick-cylindric to ovoid, to 15 cm long, to 1.5 cm thick, sessile or the lowest one pedunculate; bracts leaflike, longer than the inflorescence; pistillate scales linear-lanceolate to narrowly ovate, acuminate, sometimes awned, purplish brown with hyaline margins, as long as or longer than the perigynia; perigynia up to 150 per spike, ovoid, 4–7 mm long, pale yellow to golden brown, inflated, few-nerved, shiny, contracted to a smooth, bidentate beak 1–2 mm long; achenes trigonous, 2.0–2.2 mm long, yellow-brown, substipitate, continuous with the twisted or bent style; stigmas 3. May–June.

81. *Carex typhina.*

a. Habit.
b. Pistillate scale.

c. Perigynium.
d. Achene.

82. *Carex utriculata.*

a. Habit.
b. Pistillate scale.

c. Perigynium.
d. Achene.

Marshes, bogs.

IA, IL, IN, OH (not listed by the U.S. Fish and Wildlife Service), but OBL in all other regions.

Beaked sedge.

Carex utriculata is very similar to *C. rostrata*, but differs by its broader leaves and longer awned, pistillate scales. It differs from *C. retrorsa* in the lack of reflexed perigynia. From *C. tuckermanii*, *C. oligosperma*, and *C. vesicaria* it differs by its spongy-based culms and smooth or only slightly scabrous culms.

81. **Carex vesicaria** L. var. **monile** (Tuckerm.) Fern. Rhodora 3:53. 1901. Fig. 83.
Carex monile Tuckerm. Enum. Caric. 20. 1843.

Plants perennial, densely cespitose, with short, stout rhizomes; culms to 1 m tall, triangular, slender, usually scabrous beneath the inflorescence, purplish at the base; sterile shoots present; leaves up to 40 cm long, 4–7 mm wide, flat, septate-nodulose, green, scabrous along the margins; sheaths yellow-brown, septate-nodulose, the ligule longer than wide; staminate spikes 2–4, to 4 cm long, to 4 mm thick, well above the pistillate spikes; staminate scales acicular, yellow-brown; pistillate spikes 1–4, to 7 cm long, to 1.5 cm thick, sessile or short-pedunculate; bracts leaflike; pistillate scales lanceolate to ovate-lanceolate, acuminate, sometimes awned, yellow-brown to red-brown with hyaline margins, reaching the base of the beak of the perigynium; perigynia up to 100 per spike, globose-ovoid, 5.5–8.0 mm long, 3.5–4.0 mm broad, 1/2 to 2/3 as thick as long, inflated, membranous, yellow-green or golden brown, shiny, strongly nerved, rather abruptly tapered to a smooth, bidentate beak 1.8–2.0 mm long; achenes trigonous, 2.2–2.5 mm long, 1.7–2.0 mm wide, yellowish, substipitate, continuous with the flexuous or bent style; stigmas 3. May–August.

Depressions in swamps, wet meadows.

IL, MO, OH (OBL).

Bladder sedge.

This plant is recognized by its dense clumps, its several staminate spikes, and its few pistillate spikes with inflated perigynia.

82. **Carex viridula** Michx. Fl. Bor. Am. 2:170. 1803. Fig. 84.
Carex oederi Retz. f. *intermedia* Dudley, Bull. Cornell Univ. 2:117. 1886.
Carex flava L. var. *viridula* (Michx.) L. H. Bailey, Mem. Torrey Club 1:31. 1889.
Carex oederi Retz. var. *pumila* Fern. Rhodora 8:201. 1906.
Carex viridula Michx. f. *intermedia* (Dudley) Herrm. in Deam, Fl. Indiana 256. 1940.

Plants perennial, cespitose, with short rhizomes, sometimes mat-forming; culms to 50 cm tall, wiry, triangular, smooth, pale brown at the base; sterile shoots sometimes present; leaves up to 3.5 mm wide, dull green, sometimes folded, rather stiff, slightly scabrous along the margins; sheaths green-nerved, with a broad, loose, hyaline ventral band; terminal spike staminate but often with a few perigynia at the tip, rarely nearly entirely pistillate; staminate scales obtuse, reddish brown with a green center and hyaline margins; pistillate spikes up to 6, broadly ovoid to short-cylindric, crowded near the summit of the culm, up to 1.5 cm long, 2–3 mm thick;

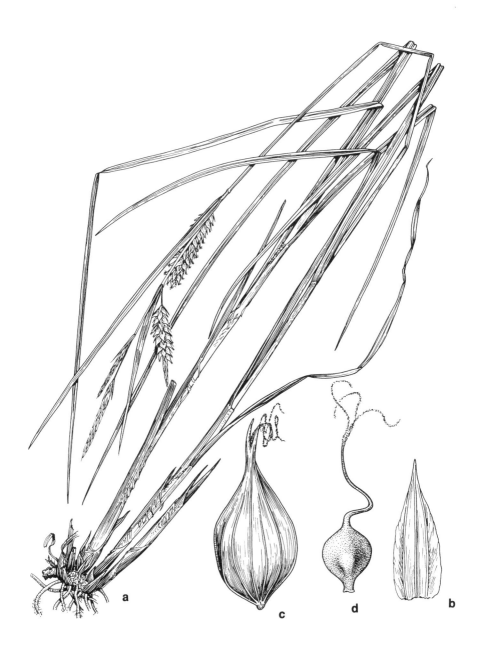

83. *Carex vesicaria*
var. *monile*.

a. Habit.
b. Pistillate scale.

c. Perigynium.
d. Achene.

84. *Carex viridula.*

a. Habit.
b. Pistillate scale.
c. Perigynium.
d. Achene.

c b d a

pistillate scales obtuse, reddish brown with a green center and hyaline margins, about as long as the body of the perigynia; perigynia up to 30 per spike, 2–3 mm long, ovoid, green or yellow-green, horizontally spreading, or the lowest ones reflexed, few-nerved, with a minutely dentate beak about a third as long as the body; achenes trigonous, with concave sides, black, shiny, 1.2–1.3 mm long, substipitate; stigmas 3. May–September.

Marshes, around lakes and ponds, wet meadows, pannes, fens, seeps.

IA, IL, IN, MO, OH (OBL).

Greenish sedge.

This species is somewhat similar to *C. cryptolepis* in having some of the lowest perigynia in a spike reflexed. It differs in its more cylindrical pistillate spikes and the beak of the perigynium that is only about one-third as long as the body.

Specimens with the terminal spike nearly entirely pistillate may be called f. *intermedia*.

83. **Carex vulpinoidea** Michx. Fl. Bor. Am. 2:169. 1803. Fig. 85.
Carex multiflora Muhl. ex Willd. Sp. Pl. 4:243. 1805.
Carex setacea Dewey, Am. Journ. Sci. 9:61. 1825.
Carex scabrior Sartw. in Boott, Illustr. Carex 3:125. 1862.
Carex xanthocarpa Bicknell var. *annectens* Bicknell, Bull. Torrey Club 23:22. 1896.
Carex annectens (Bicknell) Bicknell, Bull. Torrey Club 35:492. 1908.

Plants perennial, densely cespitose, from fibrous roots and short, stout, dark rhizomes; culms erect, up to 1 m tall, rough to the touch, at least beneath the inflorescence, dark brown at the base; leaves 3–several, up to 1.2 m long, 2–6 mm wide, flat or slightly canaliculate, rough along the margins, the upper ones exceeding the culms; sheaths open, tight, green-nerved, septate, with a broad, opaque, russet-maculate, often septate-nodulose ventral band, convex at the mouth; ligule broader than long; inflorescence composed of many spikes in an elongated, interrupted head up to 10 cm long and up to 1.5 cm broad; bracts setaceous, subtending most of the spikes, up to 5 cm long; spikes many, the staminate flowers above the pistillate; pistillate scales lanceolate, awned, hyaline to greenish brown and with a green midnerve, as long as or longer than the perigynia; perigynia several per spike, ovoid-lanceoloid to ovoid, 2–3 mm long, 1.2–1.8 mm broad, plano-convex, green to stramineous, nerveless or 2- to 4-nerved on the dorsal face, with a narrow, corky margin, tapering to a 2-toothed, serrulate or smooth beak about 0.8–1.2 mm long; achenes 1.2–1.4 mm long, lenticular, ovoid, red-brown, glossy, apiculate, jointed to the style, the style swollen at base; stigmas 2, reddish brown. May–August.

Swamps, wet meadows, moist open ground.

IA, IL, IN, KS, KY, MO, NE, OH (OBL).

Foxtail sedge.

This species and *C. brachyglossa* are distinguished by their long, slender, uninterrupted inflorescences with conspicuous setaceous bracts subtending most of the spikes. *Carex vulpinoidea* differs from *C. brachyglossa* in its greenish or stramineous, narrower perigynia with beaks more than 0.7 mm long, and its leaves usually as long as or longer than the culm.

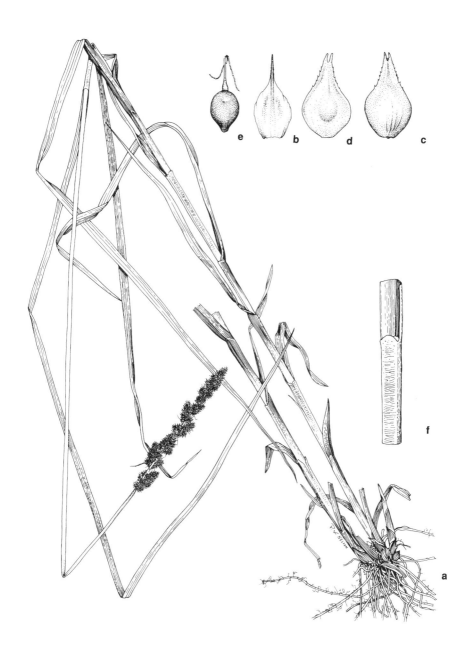

85. *Carex vulpinoidea.*
a. Habit.
b. Pistillate scale.

c. Perigynium, dorsal view.
d. Perigynium, ventral view.
e. Achene.

f. Sheath.

3. **Cladium** P. Browne—Twig Sedge

Usually rather coarse, rhizomatous or stoloniferous perennials; leaves flat or channeled; inflorescence paniculate, with numerous rays bearing umbelliform cymes of fascicles of spikelets; spikelets with several loosely imbricated scales, the lower ones empty, the next one or two above usually staminate, the terminal one perfect; bristles none; stamens 2; style 2- or 3-cleft, not persistent as a beak on the achene; ovary superior, 1-locular; achenes ovoid or globular.

Cladium is composed of about fifty species found in most parts of the world. Only the following occurs in the central Midwest.

1. **Cladium mariscoides** (Muhl.) Torr. Ann. Lyc. N.Y. 3:372. 1836. Fig. 86.
Schoenus mariscoides Muhl. Gram. 4. 1817.

Stoloniferous perennial with a triangular culm to nearly 1 m tall; leaves channeled, 1–3 mm wide, scaberulous on the margins; inflorescence paniculate, to 30 cm long; spikelets in clusters of 3–10, some pedunculate, others sessile; lowest scales sterile, the median ones staminate, the terminal one perfect; stamens 2; style usually 2-cleft; bristles 0; achene mitriform, truncate at base, apiculate at apex, 1.3–1.8 mm long. July–September.

Swamps, bogs.

IL, IN, KY, OH (OBL).

Twig sedge.

This species has the appearance of a *Rhynchospora*, but it lacks the tubercle on the achene and lacks bristles that subtend the achene. The lower part of each leaf is flat, while the upper part in inrolled.

4. **Cyperus** L.—Flatsedge

Cespitose annuals or rhizomatous perennials, sometimes bearing tubers; leaves flat or involute, glabrous or sometimes scabrous on the margins, surpassing or shorter than the culms; inflorescence composed of 1–several sessile, compact heads and frequently with 1–several simple or compound rays, subtended by 1–several involucral bracts; spikes with 5–numerous spikelets radiating in all directions, horizontally spreading, ascending, or reflexed; spikelets mostly flat, 2- to 40-flowered; scales 2-ranked, sometimes outwardly recurved at the tip, obscurely or conspicuously nerved; stamens 1–3; styles 2- or 3-cleft; achenes lenticular or trigonous, sometimes wrinkled or minutely pebbled.

As treated here, *Cyperus* consists of about 350 species distributed worldwide. Only the following may sometimes occur in aquatic situations in the central Midwest. *Cyperus* and *Dulichium* are the only genera of Cyperaceae with flattened spikelets. In *Dulichium*, the spikelets are axillary.

1. Stigmas 2; achenes lenticular.
  2. Scales straw-colored.
    3. Scales slightly recurved at tip; spikelets up to 1.9 mm wide .... 11. *C. polystachyos*
    3. Scales appressed at tip; spikelets 2 mm wide or wider.
      4. Achenes black, nearly as broad as long, with transverse wrinkles ...... 8. *C. flavescens*
      4. Achenes drab or gray, longer than broad, without transverse wrinkles but finely
        pebbled .... 2. *C. bipartitus*

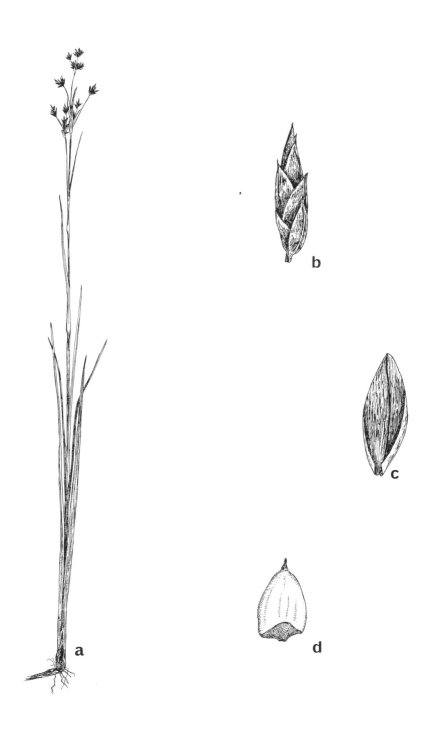

86. *Cladium mariscoides.*   a. Habit.
                            b. Spikelet.

c. Scale.
d. Achene.

2. Scales suffused with red-brown or purple.
 5. Scales with conspicuous white margins, the tips somewhat spreading; achenes black . ... 9. *C. flavicomus*
 5. Scales without white margins, the tips appressed; achenes drab or gray.
  6. Styles divided nearly to base, persistent and conspicuously exserted to 4 mm from the scales .... 4. *C. diandrus*
  6. Styles divided to about the middle, falling away early, hidden by the scales or exserted to 2 mm from the scales .... 2. *C. bipartitus*
1. Stigmas 3; achenes trigonous.
 7. Scales with strongly recurved tips .... 13. *C. squarrosus*
 7. Scales with tips either appressed or only slightly spreading.
  8. Clusters of spikelets spherical or globose, with spikelets radiating in all directions.
   9. Annuals up to 35 cm tall; one bract subtending the inflorescence strictly erect; achenes ellipsoid to oblongoid, 0.5–1.0 mm long .... 1. *C. acuminatus*
   9. Perennials at least 35 cm tall; all bracts subtending the inflorescence spreading to ascending, but never strictly erect; achenes linear, 1.0–1.4 mm long .... 12. *C. pseudovegetus*
  8. Clusters of spikelets hemispherical, cylindrical, ellipsoid, or lanceoloid, but not spherical or globose.
   10. Spikelets all arising from nearly the same point on the axis .... 3. *C. dentatus*
   10. Spikelets arising from either side of an elongated axis.
    11. Scales 1.0–1.5 mm long; achenes 0.9–1.0 mm long; roots usually reddish .... 6. *C. erythrorhizos*
    11. Scales 1.5–4.5 mm long; achenes 1.0–2.8 mm long; roots not reddish.
     12. Scales remote, the tip of one just reaching the base of the one above, giving the spikelet a zigzag appearance .... 5. *C. engelmannii*
     12. Scales approximate and overlapping, the spikelets never appearing zigzag.
      13. Rhizomes scaly and usually ending in a tuber; scales at the tips of the spikelets slightly spreading, giving the spikelet a serrated margin and an obtuse apex .... 7. *C. esculentus*
      13. Rhizomes absent or merely becoming hard and cormlike; scales at the tips of the spikelets appressed, giving the spikelet a smooth margin and a pointed apex.
       14. Plants annual, without rhizomes; scales ferruginous or golden brown, 1.7–3.0 mm long; achenes obovoid-oblongoid .... 10. *C. odoratus*
       14. Plants perennial, with hardened bases; scales straw-colored, 3.5–4.5 mm long; achenes linear .... 14. *C. strigosus*

1. Cyperus acuminatus Torr. & Hook. Ann. Lyc. N.Y. 3:435. 1836. Fig. 87.

Cespitose annual with fibrous roots; culms 4–35 cm tall, usually stramineous; leaves few, 1–2 mm wide, nearly equaling or slightly exceeding the culm; involucral bracts 2–4, exceeding the inflorescence, one of them strictly erect, the longest to 8 cm long, 1–2 mm wide; inflorescence of numerous spikes 1–2 cm in diameter, borne on rays 1–4 cm long, usually with one or more sessile spikes; spikelets flattened, ovate to oblong, rarely linear, ascending or spreading, closely imbricated, 12- to 40-flowered; scales ovate, acute, outwardly curved at the tip, 1.8–2.6 mm long, pale, with one conspicuous central nerve and usually 2–4 very obscure lateral ones;

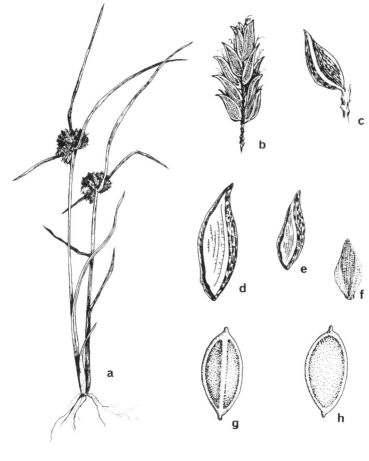

87. *Cyperus acuminatus.*

a. Habit.
b. Spikelet.

c, d, e, f. Scales.
g, h. Achenes.

stamen 1; style 2-cleft; achenes ellipsoid to oblongoid, pointed at either end, 0.5–1.0 mm long, 0.5 mm broad, much surpassed by the scales, stramineous. June–October.

Around ponds and lakes, along streams and spring branches, moist upland areas. IA, IL, IN, KS, KY, NE, MO, OH (OBL).

Acuminate flatsedge.

The crowded, hemispherical inflorescence subtended by a strictly erect bract distinguishes this rather low-growing species. The tips of the scales are usually slightly recurved.

2. **Cyperus bipartitus** Torr. Ann. Lyc. N.Y. 3:257. 1836. Fig. 88.
Cyperus diandrus Torr. var. *castaneus* Torr. Ann. Lyc. N.Y. 3:2252. 1836.
Cyperus rivularis Kunth, Enum. 2:6. 1837.
Cyperus rivularis Kunth f. *elongatus* Boeckl. Linnaea 35:4553. 1868.

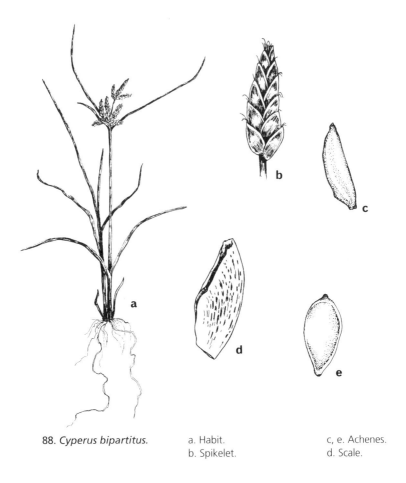

88. *Cyperus bipartitus.*   a. Habit.   c, e. Achenes.
                              b. Spikelet.   d. Scale.

Cyperus rivularis Kunth var. *elutus* C. B. Clarke, Journ. Linn. Soc. 21:65. 1884.
Cyperus rivularis Kunth f. *elutus* (C. B. Clarke) Kükenth. Pflanzenr. 4:383. 1936.
Cyperus bipartitus Torr. f. *elongatus* (Boeckl.) Mohlenbr. Sedges: Cyperus to Scleria, ed. 2, 184. 2001.
Cyperus bipartitus Torr. f. *elutus* (C. B. Clarke) Mohlenbr. Sedges: Cyperus to Scleria, ed. 2, 184. 2001.

Annual from fibrous roots; culms to 50 cm tall, smooth; leaves to 3 mm broad, nearly equaling the culm, smooth; inflorescence of 1 or 2 sessile heads and usually with 1–5 rays up to 8 cm long, with 3 (–4) involucral bracts much exceeding the inflorescence; spikes with up to 10 spikelets radiating in all directions; spikelets very flat, blunt, 8- to 35-flowered, to 25 mm long, to 4 mm broad; scales closely imbricated, ovate, obtuse, 2.0–2.5 mm long, strongly colored red-brown from near the midvein to the margins, or rarely stramineous; stamens 2 or rarely 3; styles 2-cleft to about the middle, not exserted or exserted only about 2 mm beyond the scaales, early deciduous; achenes narrowly obovoid, 1.0–1.5 mm long, 0.5–0.8 mm broad, chestnut or grayish. June–October.

Around ponds and lakes, along streams and spring branches, fens, sometimes in shallow water.

IA, IL, IN, KY, MO, OH (FACW+), KS, NE (FACW).

River flatsedge.

The scales of the spikelets are usually suffused with red-brown or purple. These plants differ from the similar appearing *C. diandrus*, another species with red-brown scales, in that the styles of *C. bipartitus* are shorter and early deciduous and are not conspicuous. *Cyperus bipartitus* f. *elutus*, a form that is devoid of red-brown pigmentation in its scales, resembles *C. flavescens* but may be distinguished by its drab or gray achenes that lack horizontal wrinkles. Plants with culms longer than 30 cm have sometimes been called f. *elongatus*.

The U.S. Fish and Wildlife Service calls this species *C. rivularis*.

3. **Cyperus dentatus** Torr. Fl. N. Mid. U.S. 1:61. 1824. Fig. 89.

Perennial with tuber-bearing rhizomes; culms triangular, slender, to 50 cm tall; leaves numerous, to 5 mm wide, the largest ones about reaching the summit of the plant; inflorescence more or less crowded into a hemispherical cluster, the branches all arising from near the same point, subtended by several bracts; spikes short, hemi-spherical, up to 3 cm across; spikelets flattened, few to many, up to 15 mm long, 8- to 25-flowered; scales ovate, prolonged into a short point at the tip, 2–3 mm long; styles 3-cleft; achenes obovoid-trigonous, 0.8–1.0 mm long, 0.3–0.5 mm wide. June–September.

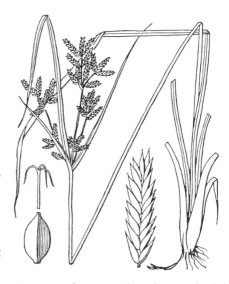

89. *Cyperus dentatus*. Habit, achene, and spikelet.

Along shores, rarely in shallow water.

IN (FACW+).

Toothed flatsedge.

This coastal plain species is distinguished by its hemispherical cluster of spikelets that all arise from the same point in the inflorescence.

4. **Cyperus diandrus** Torr. Cat. Pl. N.Y. 90. 1819. Fig. 90.

Annual from fibrous roots; culms to 40 cm tall, smooth; leaves to 3 mm wide, nearly equaling the culms, smooth; inflorescence of 1 or 2 sessile heads and usually 1–5 rays up to 6 cm long, with 3 involucral bracts exceeding the inflorescence; spikes with up to 10 spikelets radiating in all directions; spikelets very flat, blunt, 5- to 35-flowered, to 25 mm long, to 4 mm broad; scales closely imbricated, ovate, 2–3 mm long, obtuse, with a reddish purple band along the margins, which sometimes

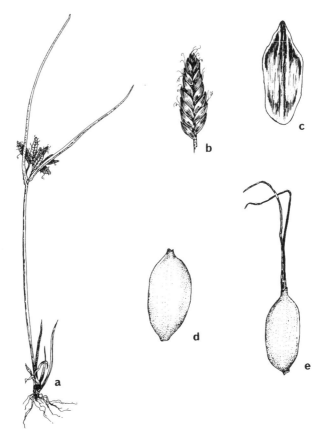

90. *Cyperus diandrus.*

a. Habit.
b. Spikelet.

c. Scale.
d, e. Achenes.

extends to near the midrib; stamens 2 or rarely 3; styles 2-cleft nearly to the base, protruding to 4 mm from the tips of the scales, persistent; achenes narrowly obovoid, 1.0–1.5 mm long, somewhat more than half as broad. June–October.

Sloughs, ditches, along rivers and streams, occasionally in shallow water.

IA, IL, IN, MO, NE (FACW+), KY (FACW).

Red-brown flatsedge.

Separation based on the red-brown pigmentation of the scales is not too reliable. Whether the style is cleft to the middle or to the base is difficult to determine, frequently due to the fragility of the styles. The most easily observed difference is that the styles of *C. diandrus* project to 4 mm from the scales and are persistent; the styles of *C. bipartitus* are included within the scales or project to only 2 mm and are rarely persistent.

5. **Cyperus engelmannii** Steud. Syn. Pl. Cyp. 47. 1855. Fig. 91.
Cyperus ferax L. C. Rich. ssp. *engelmannii* (Steud.) Kükenth. Pflanzenr. 20:620. 1936.

Rather coarse annual with fibrous roots; culms to 60 cm tall, smooth; leaves to 6 mm wide, usually exceeding the culm, smooth; inflorescence of 1–several sessile

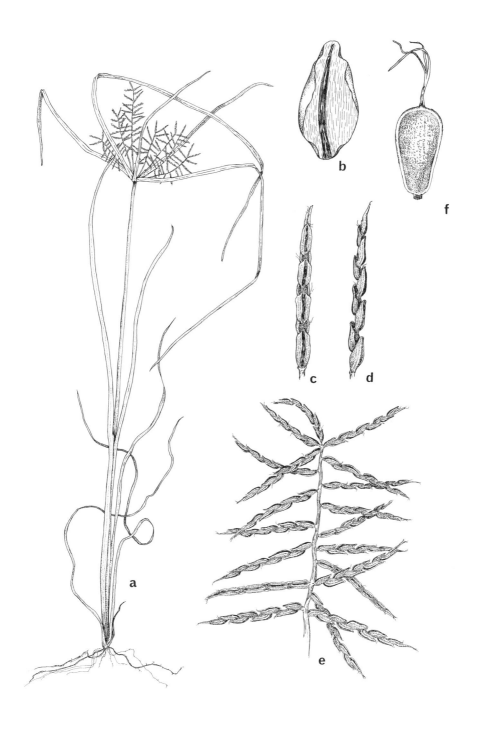

91. *Cyperus engelmannii.*
a. Habit.

b. Scale.

c, d. Spikelets.

e. Inflorescence.

f. Achene.

spikes and 1–5 rays, with up to 6 involucral bracts much surpassing the inflorescence; spikes ellipsoid to oblongoid, with numerous horizontally radiating or ascending spikelets; spikelets terete, slender, to 18-flowered, 10–20 mm long, the flowers remote; scales brown, or reddish, ovate-lanceolate, acute, 2.2–3.0 mm long, with 7–13 rather faint nerves, the tip of one barely reaching the base of the one above it on the same side, thereby giving the spikelet a rather zigzag appearance; rachilla winged; achenes linear-oblong, 1.5–2.2 mm long. July–October.

Wet ground, occasionally in shallow standing water.

IA, IL, IN, KY, MO, NE, OH. The U.S. Fish and Wildlife Service considers this plant to be a variation of *C. odoratus*. *Cyperus odoratus* is FACW.

Engelmann's flatsedge.

The scales of the spikelets, which barely reach the base of the scale above, giving the spikelet a distinctly zigzag appearance, are diagnostic for this species and clearly distinguish it as a separate species.

6. **Cyperus erythrorhizos** Muhl. Descr. Gram. 20. 1817. Fig. 92.
Cyperus halei Torr. ex Britt. Bull. Torrey Club 13:213. 1886.
Cyperus erythrorhizos Muhl. var. *halei* (Torr.) Kükenth. in Kükenth. Pflanzenr. 20:59. 1936.

Annual from fibrous, usually reddish, roots; culms to 1.3 m tall, occasionally dwarfed to 1 cm tall, smooth; leaves to 10 mm wide, shorter than to equaling the culm, with scabrous margins and with the lower sheaths usually purplish near the base; inflorescence of 1–several sessile spikes and numerous simple or compound rays, with up to 8 involucral bracts, most of which surpass the inflorescence; spikes cylindric with numerous horizontally spreading or ascending spikelets; spikelets falcate, 6- to 36-flowered, 3–20 mm long, with closely imbricated scales; scales reddish brown with stramineous margins, rarely stramineous throughout, with a green midrib, broadly lanceolate, mucronate, faintly nerved, 1.0–1.5 mm long; stamens 2 or 3; style 3-cleft nearly to the middle; rachilla winged; achene trigonous, ovoid, gray or whitish, 0.8–1.0 mm long, about 0.5 mm wide. July–October.

Sloughs, around ponds and lakes, mud flats, sand bars, occasionally in shallow water.
IA, IL, IN, KS, MO, NE (OBL), KY, OH (FACW+).
Red-rooted flatsedge.

This species is distinguished by its very small scales of the spikelets and the usually reddish roots. Plants as small as one centimeter tall may be found flowering.

7. **Cyperus esculentus** L. Sp. Pl. 1:45. 1753. Fig. 93.
Cyperus phymatodes Muhl. Descr. Gram. 23:1817 (in part).
Cyperus esculentus L. var. *leptostachyus* Boekl. Linn. 36:290. 1870.
Cyperus esculentus L. f. *angustispicatus* (Britt.) Fern. Rhodora 44:141. 1942.

Perennial from numerous conspicuously scaly rhizomes terminating in a small hard tuber; culms rather stout, to 1 m tall, smooth; leaves flat, to 10 mm wide; inflorescence with 1–several sessile spikes and 1–10 rays, with 3–10 broad involucral bracts surpassing the inflorescence; spikes mostly cylindric, with numerous horizontally radiating or ascending spikelets; spikelets flattened, 6- to 30-flowered, 6–35 mm long, 1.5–3.0 mm broad; scales loosely imbricate or spreading somewhat,

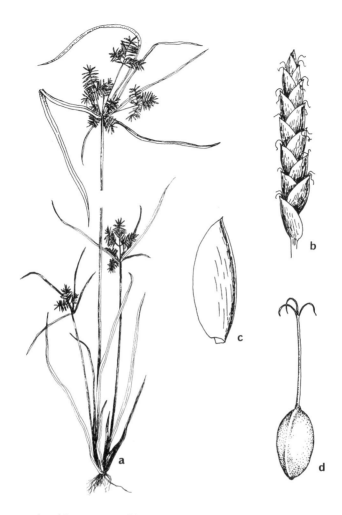

92. *Cyperus erythrorhizos.* a. Habit.   c. Scale.
                              b. Spikelet.   d. Achene.

ovate, obtuse to acute or rarely with the terminal scale acuminate, 7- to 11-nerved,
2–3 mm long, stramineous or golden brown, scarious at the tip; rachilla winged;
achenes oblongoid to narrowly oblongoid, 1.2–1.8 mm long, 0.4–0.8 mm broad,
brownish or grayish. June–October.

 Along rivers and streams, around ponds and lakes; also a weed in cultivated soils.
IA, IL, IN, KS, KY, NE, MO, OH (FACW).

 Yellow nut sedge; chufa nut.

 This aggressive species propagates readily by its underground tubers. The tubers
are edible and rooted out of the ground and eaten. Robust plants with spikelets 20–
35 mm long and more loosely arranged scales may be designated as var. *leptostachyus.*
Cyperus esculentus is usually not found in standing water.

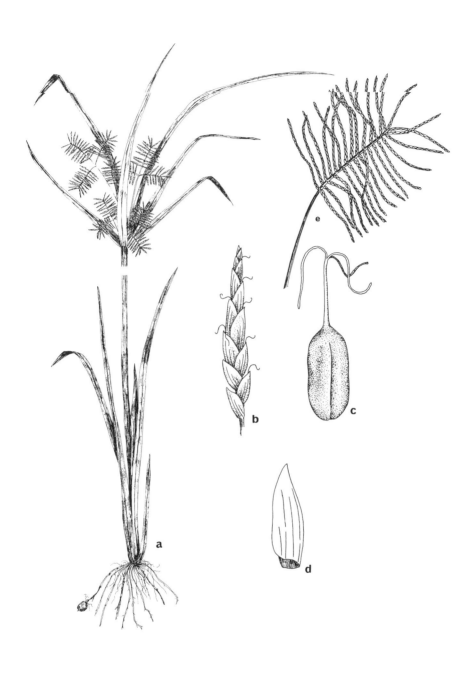

93. *Cyperus esculentus.*
a. Habit.
b. Spikelet.

c. Achene.
d. Scale.

e. Inflorescence of var.
leptostachyus.

94. *Cyperus flavescens.* a. Habit.   c, d. Scales.
                                             b. Spikelet.   e. Achene.

8. **Cyperus flavescens** L. Sp. Pl. 1:46. 1753. Fig. 94.
Cyperus poaeformis Pursh, Fl. Am. Sept. 1:50. 1814.
Cyperus flavescens L. var. *poaeformis* (Pursh) Fern. Rhodora 41:529. 1939.

Cespitose annual with fibrous roots; culms 8–45 cm tall, straw-colored, rarely becoming purplish near the base, smooth, 0.5–1.5 mm broad; leaves 0.5–2.0 mm wide, about 1/2 to 2/3 as long as the culms; bracts usually 3, rarely 2 or 4, at least one and sometimes all of them exceeding the inflorescence, the longest to 12 cm long, 0.5–2.0 (rarely to 3.0) mm broad; spikes crowded into condensed umbels or occasionally with 1–3 rays up to 3 cm long; spikelets very flat, 12- to 25-flowered, 5–15 mm long, 2–3 mm broad, obtuse at the tip; scales pale yellowish brown, ovate, obtuse, with hyaline margins, distinctly green-keeled, 1.5–2.5 mm long; stamens 3; styles deeply 2-cleft; achenes flattened, obovoid, with minute transverse wrinkles, 0.8–1.0 mm long, nearly as broad, apiculate, black. July–October.

Fens, along streams and spring branches, around lakes and ponds, sometimes in standing water.

IA, IL, IN, KS, KY, MO, OH (OBL).

Yellowish flatsedge.

This annual strongly resembles *C. bipartitus* and *C. diandrus* but lacks the red-brown to purplish scales of these two species. It is very similar to the non-purple form of *C. bipartitus* but may be distinguished by its black achenes with minutely transverse wrinkles.

9. **Cyperus flavicomus** Michx. Fl. Bor. Am. 1:27–28. 1803. Fig. 95.
Pycreus albomarginatus Mart. & Schrad. ex Nees, Fl. Bras. 2 (1):9. 1842.
Cyperus albomarginatus (Mart. & Schrad. ex Nees) Steud. Syn. Pl. Glumac. 2:10. 1854.

Annual with tufted roots; stems trigonous, to 70 cm tall; leaves up to 7 mm wide, sometimes folded; inflorescence of usually 1–3 sessile spikes and 2–8 spikes on rays, subtended by 3–7 bracts longer than the inflorescence; spikes to 3 cm long, with 6–60 spreading spikelets; spikelets up to 20 mm long, 2–3 mm wide, with 8–24 florets, the scales 1.3–1.6 mm long, narrowly ovate, slightly spreading at the obtuse tip, red-brown with a conspicuous white margin; stigmas 2; achenes 1.2–1.6 mm long, obovate, black, finely pebbled, shiny. August–October.

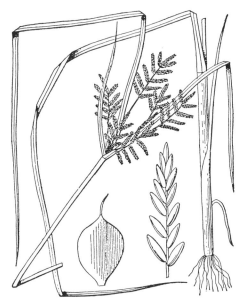

95. *Cyperus flavicomus*. Habit, achene, and spikelet.

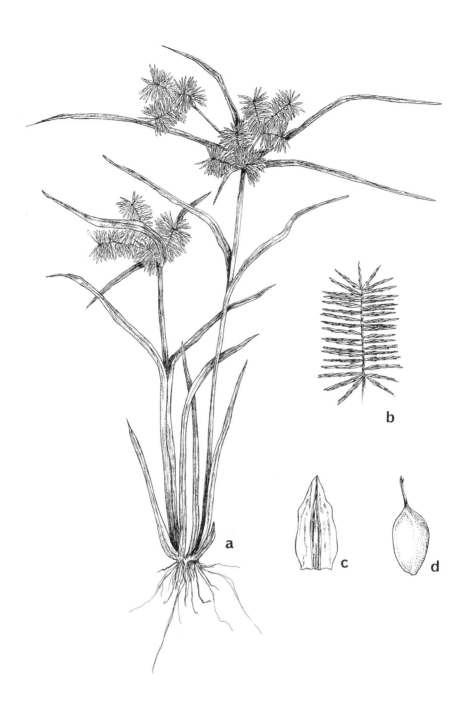

96. *Cyperus odoratus.*

a. Habit.
b. Head.

c. Scale.
d. Achene.

Marshes and crop fields, occasionally in standing water.

MO (not listed for region 3), where it grows as an emergent aquatic.

White-bordered flatsedge.

Cyperus flavicomus is the only *Cyperus* with a broad white margin on each scale of the spikelets. Traditionally in the United States, this species has been called *Cyperus albomarginatus*, the name used by the U.S. Fish and Wildlife Service.

10. **Cyperus odoratus** L. Sp. Pl. 1:46. 1753. Fig. 96.
Cyperus speciosus Vahl, Enum. 2:364. 1806.
Cyperus ferruginescens Boeckl. Linn. 36:396. 1870.
Cyperus speciosus Vahl var. *squarrosus* Britt. Bull. Torrey Club 13:214. 1886.
Cyperus ferax L.C. Rich ssp. *speciosus* (Vahl) Kükenth. var. *squarrosus* (Britt.) Kükenth. Pflanzenr. 20:620. 1936.

Coarse annual with fibrous roots; culms to 1 m tall, smooth; leaves flat, usually not equaling the culm, to 12 mm broad; inflorescence with 1–2 sessile heads and 2–12 simple or compound rays, with several of the numerous involucral bracts surpassing the inflorescence; spikes cylindric, with numerous horizontally radiating or ascending spikelets; spikelets flattened, 8- to 25-flowered, to 25 mm long and with appressed or rarely loosely spreading scales; scales ovate, 1.7–3.0 mm long, with numerous faint nerves, reddish brown or golden brown, the terminal scale often subulate; rachilla winged; achenes oblong-obovoid, 1.0–1.7 mm long, red-brown or brown. July–October.

Along rivers and streams, around ponds and lakes, floodplains, sandbars, occasionally in shallow water.

IA, IL, IN, KS, KY, MO, NE, OH (FACW).

Rusty flatsedge.

The nomenclature for this species has been in a state of confusion for years. If Linnaeus's *Cyperus odoratus* from Europe is distinct from North American plants, then our species becomes *C. ferruginescens*.

This species is subjected to frequent inundations, which give rise to peculiar individuals 1–20 cm tall with spikelets to 25 mm long and scales 2.5–3.0 mm long and very loosely spreading. The subulate terminal scales of this species help distinguish it from *C. esculentus*.

11. **Cyperus polystachyos** Rottb. var. **texensis** (Torr.) Fern. Rhodora 41:530. 1939. Fig. 97.
Cyperus microdontus Torr. var. *texensis* Torr. Ann. Lyc. N. Y. 3:430. 1836.

Annual with tufted roots; stems triangular, to 35 cm tall; leaves 1–3 mm wide, sometimes folded, shorter than the culms; inflorescence of 1–5 sessile spikes and 1–6 spikes on rays, subtended by 3–5 long bracts; spikes ovoid to cylindrical, to 3 cm across, with 10–30 spikelets; spikelets to 18 mm long, to 1.9 mm wide, narrowly elliptic to linear, with up to 25 florets; scales 1.5–2.3 mm long, oblong-ovate, straw-colored, with slightly recurved tips; stigmas 2; achenes 0.8–1.2 mm long, oblong, dark gray to black, finely pebbled, shiny. July–October.

Sloughs, around ponds, along spring branches, in acid seeps.

97. *Cyperus polystachyos* var. *texensis.* Habit, spikelet, achene, and scale.

KY, MO (FACW).
Texas flatsedge.
This species is found rarely in standing water. It is distinguished by its slightly recurved tips of the scales of the spikelets.

12. **Cyperus pseudovegetus** Steud. Synops. Cyper. 24. 1855. Fig. 98.
Cyperus virens Gray, Man. ed. 2, 493. 1856, non Michx. (1803).

Perennial from a short rhizome; culms 35–65 cm tall, glabrous; leaves 2–4 mm wide, usually equaling the culm, glabrous; inflorescence of 1–several sessile heads and numerous compound rays; spikes with numerous radiating spikelets; spikelets 5- to 13-flowered, 3.0–6.5 mm long, with rather loosely arranged scales; scales narrow, subacute, with slender recurved tips, 1.8–2.2 mm long, faintly nerved, pale

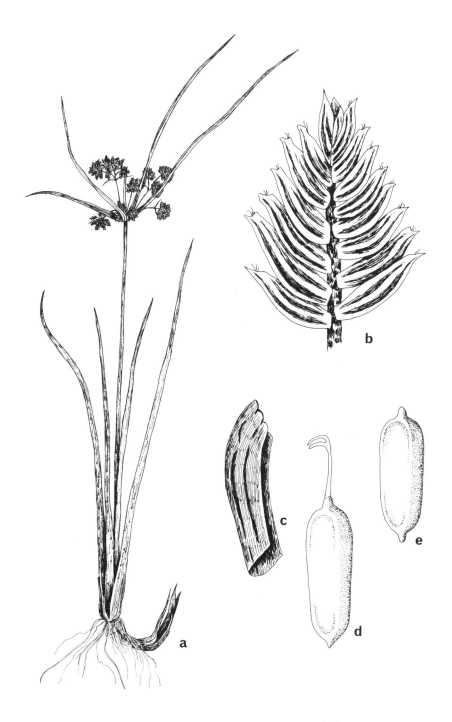

98. *Cyperus pseudovegetus.*   a. Habit.
                                     b. Spikelet.

c. Scales.
d, e. Achenes.

greenish yellow; stamen 1; style 3-cleft; achenes linear, 1.0–1.4 mm long, stipitate, stramineous. June–October.

Ditches, along streams, around ponds, sinkhole ponds, swamps, bottomland forests; also upland sites.

IL, IN, KS, KY, MO (FACW).

Flatsedge.

This species resembles *Cyperus acuminatus* but differs by its spreading involucral bracts. At least one of the involucral bracts in *C. acuminatus* is erect.

This species rarely occurs in standing water.

13. **Cyperus squarrosus** L. Cent. Pl. II, 6. 1756. Fig. 99.
Cyperus aristatus Rottb. Descr. & Icon. 6:23. 1773.
Cyperus inflexus Muhl. Desc. Gram. 16. 1817.
Cyperus aristatus Rottb. var. *inflexus* (Muhl.) Boeckl. Linnaea 35:500. 1868.

Cespitose annual with an odor of slippery elm *(Ulmus rubra)*; roots fibrous, forming a dense mat; culms 3–15 cm tall, very slender, purplish-tinged at base, smooth; leaves 2–3 on each culm, 0.5–1.5 mm wide; involucral bracts 2–4, all exceeding the inflorescence, the longest to 8 cm, 0.5–2.0 mm broad; spikes crowded into a sessile headlike cluster, occasionally with a few on rays up to 2.5 cm long; spikelets flattened, 6- to 18-flowered, 3–9 mm long, 1–2 mm wide; scales oblong to oblong-lanceolate, prominently 7- to 9-nerved, 1–2 mm long, green when young becoming reddish brown or brown at maturity; rachilla wingless; stamen 1; style 3-cleft, deciduous; achenes trigonous, obovoid, minutely pebbled, 0.5–1.0 mm long, 0.3–0.5 mm broad, pale brown. May–October.

Along rivers and streams, around lakes and ponds; also upland sites.

IA, IL, IN, KS, MO, NE, (OBL), KY, OH (FACW+).

Recurved flatsedge.

This small annual is distinguished by its very strongly recurved scales of each spikelet. The plants, particularly when dried, have a strong odor of slippery elm. The U.S. Fish and Wildlife Service calls this species *C. aristatus*.

14. **Cyperus strigosus** L. Sp. Pl. 1:47. 1753. Fig. 100.
Cyperus stenolepis Torr. Ann. Lyc. N.Y. 3:263. 1836.
Cyperus strigosus L. f. *robustior* Kunth, Enum. 2:88. 1837.
Cyperus strigosus L. var. *robustior* (Kunth) Britt. Bull. Torrey Club 13:212. 1886.
Cyperus strigosus L. var. *stenolepis* (Torr.) Kükenth. in Fedde, Rep. 23:189. 1926.

Perennial from a hard cormlike rhizome; culms to 1.2 m tall, smooth; leaves flat, to 12 mm wide, some of them surpassing the culms; inflorescence with 1–2 sessile heads and 2–12 simple or compound rays, with several of the numerous bracts much longer than the inflorescence; spikes cylindric, with numerous horizontally radiating or ascending spikelets; spikelets strongly flattened, 3- to 25-flowered, to 30 mm long, with appressed or loosely ascending scales; scales acute and often mucronulate, 3.5–5.0 mm long, 7- to 11-nerved, golden brown, with scarious margins; rachilla winged; achene linear to linear-oblongoid, 1.3–2.2 mm long, 0.4–0.7 mm broad. July–October.

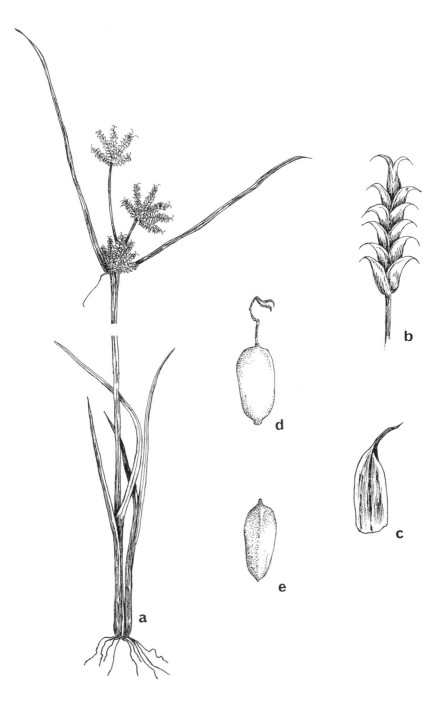

99. *Cyperus squarrosus.*   a. Habit.   c. Scale.
b. Spikelet.   d, e. Achenes.

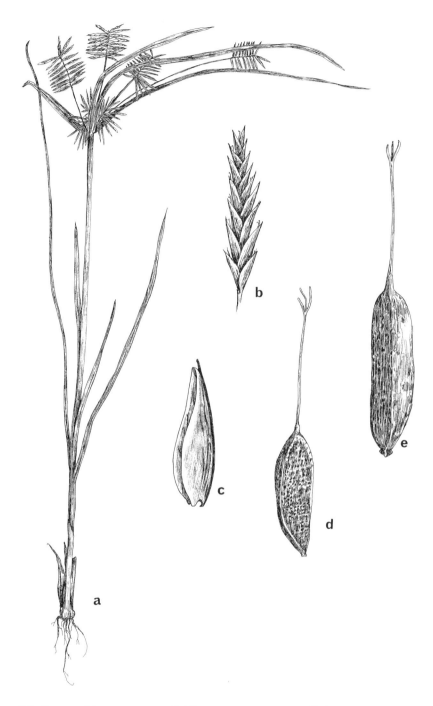

100. *Cyperus strigosus.*

a. Habit.
b. Spikelete.

c. Scale.
d, e. Achenes.

Sloughs, along rivers and streams, around ponds and lakes, ditches, sloughs.
IA, IL, IN, KS, KY, MO, NE, OH (FACW).
Straw-colored flatsedge.

Cyperus strigosus is one of our most variable species with every degree of intergradation occurring among several characters. Very robust specimens with appressed scales (f. *robustior*) or loosely arranged scales (var. *stenolepis*) occur, but intergradations of all degrees may be found from the smallest to the largest variations. It does not seem feasible to retain any of the named variations.

5. **Dulichium** Rich. ex Pers.—Three-way Sedge

Characters of the species. There is only one species of *Dulichium.*

1. **Dulichium arundinaceum** (L.) Britt. Bull. Torrey Club 21:29. 1894. Fig. 101.
Cyperus arundinaceus L. Sp. Pl. 1:44. 1753.
Cyperus spathaceus L. Syst. ed. 12. 735. 1767.
Dulichium spathaceum (L.) Pers. Syn. 1:65. 1805.

Perennial with a horizontal rhizome; culms simple, hollow, terete, jointed, leafy to the top, to 1 m tall; leaves 3-ranked, alternate, the lowest reduced to bladeless sheaths, to 10 cm long, to 8 mm broad, spreading to ascending, flat, clasping at the base, scabrous along the margins; inflorescence axillary from the leaf sheaths, each spike with 7–15 flattened, 2-ranked spikelets; scales decurrent with winglike projections on the joint below, oblong-lanceolate, 5–8 mm long, wrapped around each floret, green with a red-brown margin, with 5–9 nerves; florets perfect; perianth of 6–9 retrorsely barbed bristles; stamens 3; style 2-cleft, persistent; achenes flattened, narrowly elliptic, 2.5–3.5 mm long, with a beak about as long as the body. July–October.

Swamps, low ground in woods, sinkhole ponds, fens, sloughs, along spring branches, sometimes in shallow water.
IA, IL, IN, KY, MO, NE, OH (OBL).
Three-way sedge.
The hollow, terete, leafy stems make this species easily identifiable in the vegetative condition. This species sometimes forms dense floating mats.

6. **Eleocharis** R. Br.—Spikesedge

Annuals with fibrous roots or perennials with slender rhizomes or stolons; culms unbranched, usually terete; leaves usually reduced to sheaths (lowest sheath occasionally blade-bearing in *E. equisetoides*); spikelets borne in solitary, terminal, bractless heads 2- to several-flowered, each flower subtended by a scale; lowest 1–3 scales usually sterile; stamens (1–) 3; styles 2- or 3-cleft; achenes biconvex or trigonous, smooth, verrucose, longitudinally ridged, reticulate, puncticulate, or with trabeculae; dilated style base jointed to the achene as a persistent tubercle.

Mature achenes are essential to identify most species of *Eleocharis* positively. In any given area in the central Midwest, there are a few species that make up 95% of the populations of *Eleocharis*. These few may be tentatively identified using size of plants, diameter of stems, and whether the plant has rhizomes or tufted roots.

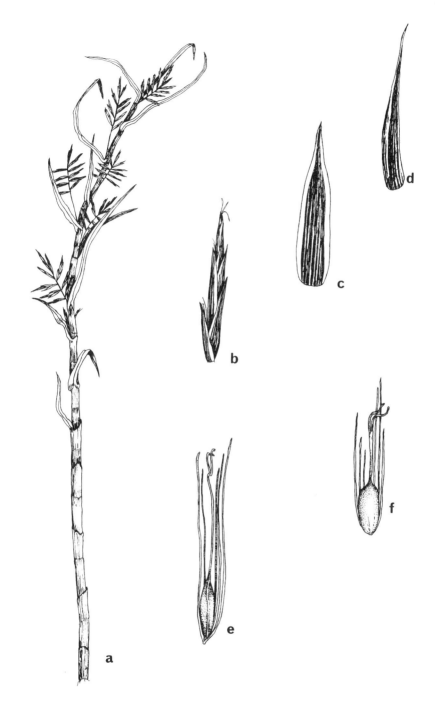

101. *Dulichium*
arundinaceum.

a. Habit.
b. Spikelet.

c, d. Scales.
e, f. Achenes.

1. Spikelet about the same thickness as the culm.
   2. Stems terete, septate by cross partitions; bristles shorter than the achene, or lacking ......
   .... 5. *E. equisetoides*
   2. Stems 3- or 4-angled, not septate; bristles slightly longer than the achene.
      3. Culms 3-angled; spikelets up to 2 cm long .... 20. *E. robbinsii*
      3. Culms 4-angled; spikelets 2–5 cm long .... 19. *E. quadrangulata*
1. Spikelet usually conspicuously thicker than the culm.
   4. Style 2-cleft (some 3-cleft styles may be intermixed); achenes obovoid-lenticular (rarely sometimes trigonous in *E. olivacea*).
      5. Plants perennial, with rhizomes.
         6. Stems strongly flattened and often twisted .... 24. *E. xyridiformis*
         6. Stems more or less terete, not flattened nor twisted.
            7. Achenes at most 1 mm long, the small, conicle tubercle 1/8–1/6 as long ....
            .... 14. *E. olivacea*
            7. Achenes averaging somewhat longer than 1 mm, the conical tubercle 1/4–1/2 as long.
               8. Basal scale 1, suborbicular, completely encircling the culm; culms averaging 0.6 mm wide .... 6. *E. erythropoda*
               8. Basal scales 1–3, never completely encircling the culm; culms averaging 1.3 mm wide.
                  9. Basal sheaths of mature culms with prominent V-shaped sinuses; culms usually somewhat soft and inflated, up to 75 cm tall .... 16. *E. palustris*
                  9. Basal sheaths of mature culms truncate to slightly oblique at apex; culms usually rigid, often more than 75 cm tall .... 10. *E. macrostachya*
      5. Plants annual, with fibrous roots.
         10. Achenes yellow to deep brown; plants usually 15–50 cm tall.
            11. Scales of spikelets pointed at tip; spikelets lanceoloid .... 9. *E. lanceolata*
            11. Scales of spikelets obtuse at tip; spikelets ovoid to oblongoid .... 15. *E. ovata*
         10. Achenes black or purple; plants usually less than 15 cm tall.
            12. Achenes up to 0.5 mm long; bristles white .... 2. *E. atropurpurea*
            12. Achenes at least 0.7 mm long; bristles brown .... 7. *E. geniculata*
   4. Style 3-cleft; achenes usually trigonous.
      13. Achenes obpyramidal, truncate at top; tubercle low, flat, closely covering entire top of achene .... 11. *E. melanocarpa*
      13. Achenes usually obovoid, rounded at top; tubercle narrower than achene or sharply differentiated from achene.
         14. Tubercle confluent with top of achene, long-conical.
            15. Achenes up to 1.5 mm long; scales up to 2.5 mm long .... 17. *E. parvula*
            15. Achenes 1.8–3.0 mm long; scales 3–8 mm long.
               16. Culms flat, to 2 mm wide; scales elliptic, obtuse; plants tufted ....
               .... 21. *E. rostellata*
               16. Culms usually 3-angled, less than 1 mm wide; scales lanceolate, acute; plants with stoloniferous rhizomes .... 18. *E. pauciflora*
         14. Tubercle conspicuously differentiated from the achene.
            17. Achenes about twice as long as wide, with 8–19 longitudinal ribs.
               18. Scales 1.5–2.2 mm long; culms 0.1–0.5 mm wide, capillary ....
               .... 1. *E. acicularis*
               18. Scales 2.5–3.0 mm long; culms about 1 mm wide, often inrolled ....
               .... 23. *E. wolfii*
            17. Achenes 2/3 to as wide as long, smooth, warty, reticulate, or with papillae.
               19. Achenes appearing smooth.

20. Tubercle subulate, 1/3–1/2 as long as the body; achenes olive or
    yellowish, 1.0–1.5 mm long .... 8. *E. intermedia*
20. Tubercle depressed, minute; achenes gray or whitish, 0.6–0.8 mm
    long .... 12. *E. microcarpa*
19. Achenes pitted or with papillae or warty or reticulate.
   21. Culms with 5 vascular bundles and conspicuously (4–) 5-angled;
      mature achenes usually olivaceous and warty .... 22. *E. verrucosa*
   21. Culms with 6–14 vascular bundles, terete or angular or compressed;
      mature achenes yellow, brown, or orange-brown, pitted or slightly warty.
     22. Culms flattened; achenes reticulate .... 3. *E. compressa*
     22. Culms angular or terete; achenes pitted or usually low-warty
       (sometimes only reticulate in *E. elliptica*).
       23. Achenes low-warty; culms usually 6- to 8-angular; plants to
         30 cm tall .... 4. *E. elliptica*
       23. Achenes pitted; culms more or less triangular or terete; plants
         up to 50 cm tall .... 13. *E. montevidensis*

1. **Eleocharis acicularis** (L.) Roem. & Schult. Syst. Veg. 2:154. 1817. Fig. 102.
Scirpus acicularis L. Sp. Pl. 1:48. 1753.
Eleocharis acicularis (L.) Roem. & Schult. var. *gracilescens* Svenson, Rhodora
31:191. 1929.

Plants commonly growing in tufts from numerous brown or reddish stolons, the
stolons spreading, about 0.3 mm thick; culms capillary, 0.1–0.5 mm thick, furrowed,
rarely terete, 0.3–2.1 dm long; sheaths red or straw-colored, often becoming free at
the apex; spikelets ovoid to lanceoloid, acute, usually flattened, 2–8 (–12) mm long;
scales reddish brown with a wide green midrib and a hyaline margin, 1.5–2.2 mm
long, ovate to lance-ovate; styles 3; achenes obovoid or ellipsoid, 0.6–0.8 mm long,
0.3–0.4 mm wide, pearly white or light brown, with 8–18 longitudinal ridges and
many close trabeculae; tubercle conical, depressed, or depressed with apiculate
center, usually with a constriction between the tubercle and achene; bristles 0–3,
very delicate when present. July–October.

Ditches, around ponds and lakes, sloughs, in springs, mud flats.

IA, IL, IN, KS, KY, MO, NE, OH (OBL).

Needle spikesedge.

This species may grow completely inundated. When it is inundated, the culms
may reach a length of 40 cm and the spikelets may be 6–12 mm long. The threadlike culms and the matted habit readily distinguish this species.

2. **Eleocharis atropurpurea** (Retz.) J. Presl & C. Presl, Reliq. Haenk. 1 (3):196.
1828. Fig. 103.
Scirpus atropurpureus Retz. Obs. Bot. 5, 14. 1789.

Tufted annual; culms to 15 cm tall, nearly terete, purple at base; spikelets 2–8 mm
long, oblongoid to ovoid, acute to obtuse at the apex, with 1, sterile, basal scale; scales
1.0–1.5 mm long, ovate, obtuse, with purple-black margins; bristles 0–5, shorter
than the achene when present; styles 2-cleft; achenes 0.5–0.6 mm long, obovoid,
more or less flattened, smooth, reddish black to black, shiny. May–October.

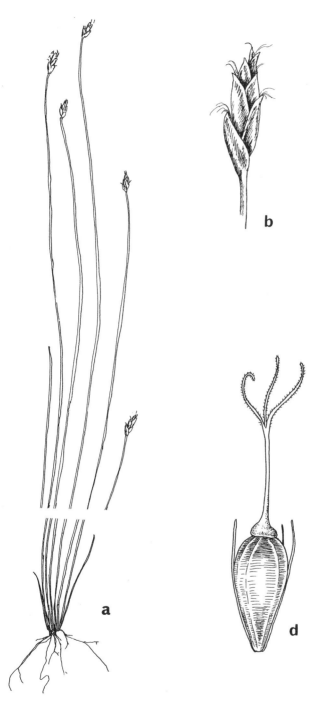

102. *Eleocharis acicularis.*
a. Habit.
b. Spikelet.
c. Scale.
d. Achene.

Around ponds, marshes, usually in prairie areas.

IA, KS, MO, NE (FACW).

Purple-black spikesedge.

The reddish black to black achenes and the scales with purple-black margins are distinctive for this species.

3. **Eleocharis compressa** Sull. Am. Journ. Sci. 42 (1):50–51. 1842. Fig. 104. *Eleocharis elliptica* Kunth var. *compressa* (Sull.) Drap. & Mohlenbr. Am. Midl. Nat. 64:365. 1960.

Perennial with rhizomes; stems to 40 cm tall, 1–2 mm in diameter, very flattened, sometimes twisted, often purple at base; spikelets 5–12 mm long, oblongoid to ellipsoid, acute to obtuse at apex, with 1 sterile, basal scale; scales 2.0–3.5 mm long, ovate, acute at apex, brown or purple-brown with a whit-

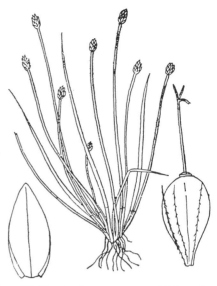

103. *Eleocharis atropurpurea*. Habit, scale, and achene.

ened tip and whitened margins; perianth bristles 0–5, shorter than the achenes when present; styles 3-cleft; achenes 1.0–1.5 mm long, obovoid, more or less trigonous, golden yellow to brown, reticulate or minutely pitted. May–July.

Ditches, along rivers, around lakes; sometimes in moist upland sites.

IA, IL, IN, KS, KY, MO, NE, OH (FACW).

Compressed spikesedge.

This species is readily recognized by its very flat, often twisted stems. The only other species with flat, twisted stems is *E. xyridiformis*, but it is a more robust plant with larger achenes and a 2-cleft style.

4. **Eleocharis elliptica** Kunth, Enum. Pl. 2:146. 1837. Fig. 105. *Eleocharis capitata* (L.) Blake var. *borealis* Svenson, Rhodora 34:200. 1932.

Perennial from straw-colored, purple, or brown rhizomes, 1–5 mm thick; sheaths straw-colored or reddish brown, closely adherent to the culm; culms terete or flat, 0.4–1.5 mm thick, 1–7 dm tall, with 6–8 vascular bundles and appearing with 6–8 low angles; spikelets ovoid and obtuse to lanceoloid and acute, 3–11 mm long; scales ovate to oblong, obtuse to acute, straw-colored to dark purple with a light margin, the apex sometimes 2-cleft; styles 3-cleft; achenes trigonous, obovoid, 0.8–1.2 mm long, reticulate or slightly warty, brown, orange-brown, or yellow; tubercle conical, depressed, or depressed with an apiculate center, all types sometimes occurring on the same plant; bristles absent. May–July.

Ditches, around lakes and ponds, along rivers and streams.

IL, IN, OH. The United States Fish and Wildlife Service does not distinguish this plant from *E. compressa* Sull. *Eleocharis compressa* is FACW.

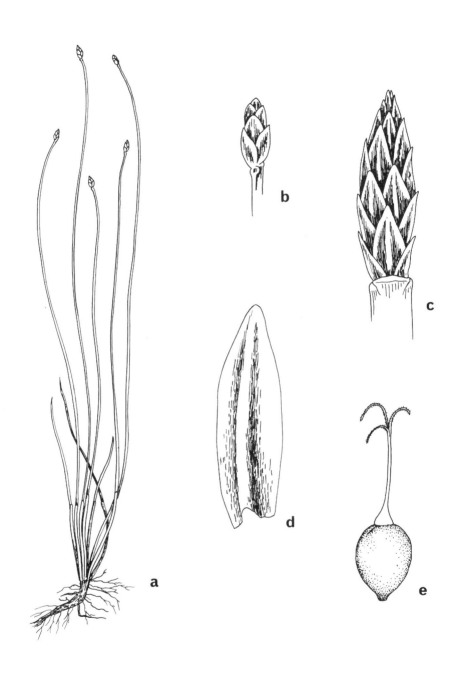

104. *Eleocharis compressa.*   a. Habit.   d. Scale.
    b, c. Spikelet.   e. Achene.

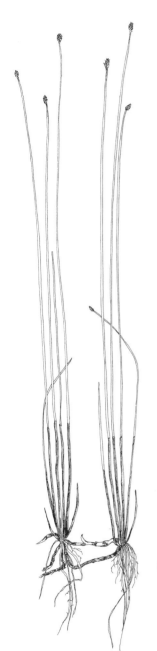

105. *Eleocharis elliptica.*　　a. Habit.　　　　　　　　d. Scale.
　　　　　　　　　　　　　　　b, c. Spikelets.

Elliptic spikesedge.

The angular culms and usually low, warty achenes distinguish this species from *E. compressa.*

5. **Eleocharis equisetoides** (Ell.) Torr. Ann. Lyc. N.Y. 3:296. 1836. Fig. 106.
Scirpus equisetoides Ell. Bot. S. C. & Ga. 1:79. 1816.

Perennial with coarse fibrous roots; culms terete, septate by cross-partitions, to 1 m tall; lowest sheaths frequently bearing a leaf; spikelets 2–4 cm long, acute, as thick as or slightly thicker than the culm; scales rather elliptic, pale, with scarious margins, to 6 mm long, to 4 mm wide; stamens 3; style 2- or 3-cleft; achenes biconvex, ovoid, 2.3–2.6 mm long, with narrowly rectangular reticulations; tubercle conical, sessile, about 1/2 as long as the achene; bristles shorter than the achene, or lacking. July–October.

Around lakes and ponds, sinkhole ponds.

IL, IN, KY, MO (OBL).

Horsetail spikesedge.

This is one of the easiest species of *Eleocharis* to recognize because of its terete, septate culms. The tip of the culm rarely may bear two spikelets.

6. **Eleocharis erythropoda** Steud. Syn. Cyp. 65. 1855. Fig. 107.
Scirpus glaucus Torr. Fl. U.S. 44. 1824, non Lam. (1791).
Eleocharis calva (Gray) Torr. Fl. N.Y. 2:346. 1843, *nomen provisiorum.*
Eleocharis palustris (L.) Roem. & Schult. var. *calva* Gray, Man. 522. 1848.

Perennials from rhizomes; culms to 70 cm tall, 0.4–1.5 mm wide, averaging 0.6 mm, red-purple at the base; spikelets ovoid to ellipsoid to lanceoloid, acute to acuminate with a cuspidate tip, 5–18 mm long, averaging 11 mm, with a large, single, suborbicular basal scale completely encircling the culm; scales ovate to lanceolate, obtuse to acute, 1.1–3.7 mm long, averaging 2.8 mm, brown or straw-colored with a light midrib and hyaline margin; achenes obovoid, 1.0–1.3 mm long, rarely 1.4–1.6 mm long (excluding the tubercle), yellow, becoming dark brown with age; tubercle narrowly conical, more than 1/2 as long as the body of the achene. May–September.

Ditches, along streams, around ponds and lakes, sloughs, sometimes in prairie areas.

IA, IL, IN, KS, KY, MO, NE, OH (OBL).

Red-based spikesedge.

This species always has culms with a red-purple base, although other species of *Eleocharis* may also be purple-based. The single scale at the base of the spikelets completely encircles the culm and is distinctive for this species.

7. **Eleocharis geniculata** (L.) Roem. & Schult. Syst. Veg. 2:150. 1817. Fig. 108.
Scirpus geniculatus L. Sp. Pl. 1:48. 1753, in part, non S. *geniculatus* L. (1762).
Scirpus caribaeus Rottb. Descr. Pl. Rar. 24. 1772, misapplied.
Scirpus geniculatus minor Vahl, Enum. 2:251. 1805.
Eleocharis dispar E. J. Hill, Bot. Gaz. 7:3. 1882.
Eleocharis capitata (L.) Blake var. *dispar* (Hill) Fern. Rhodora 8:129. 1906.

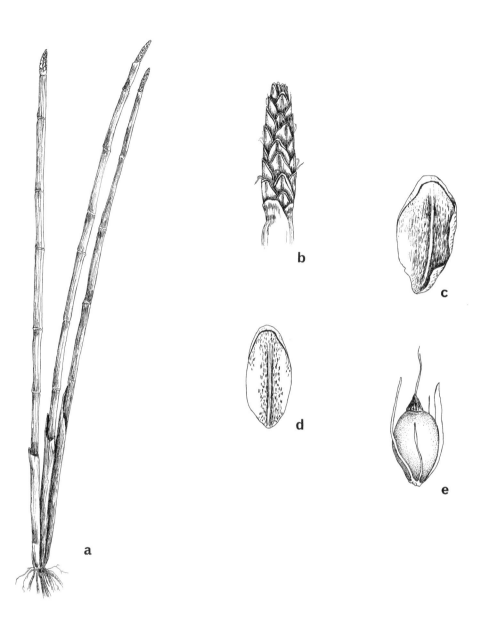

106. *Eleocharis equisetoides.* a. Habit.
　　　　　　　　　　　　　　b. Spikelet.
c, d. Scales.
e. Achene.

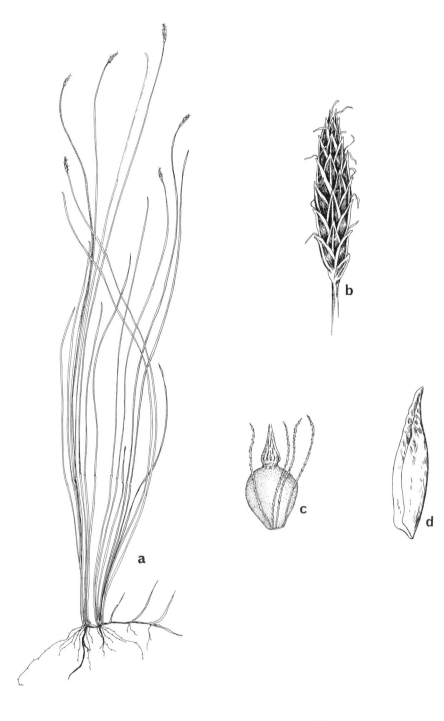

107. *Eleocharis erythropoda.*   a. Habit.
                                      b. Spikelet.

c. Achene.
d. Scale.

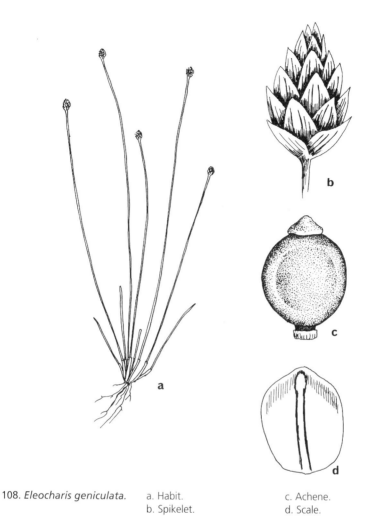

108. *Eleocharis geniculata.*   a. Habit.   c. Achene.
b. Spikelet.   d. Scale.

Eleocharis caribaea (Rottb.) Blake, Rhodora 20:24. 1918.
Eleocharis caribaea (Rott.) Blake var. *dispar* (Hill) Blake, Rhodora 20:24. 1918.

Annual from fibrous roots; culms 3.5–15.0 cm long, angled or furrowed, 0.3–0.6 mm thick; sheaths red, brown, or straw-colored, terminating at the apex in an acute, hyaline tooth; spikelets spherical in outline to ovoid, obtuse, many-flowered, 3–4 mm long, dark red with a pale midrib and hyaline margin; style 2-cleft; achenes obovoid, 0.8 mm long, 0.6 mm wide, smooth, lustrous brown, black, or purple; tubercle short, depressed or triangular; bristles 5–7, rough, usually persisting, about as long as the achene. June–September.

Around ponds and lakes, not usually in standing water.

IL, IN, KS (FACW), KY (OBL).

Jointed spikesedge.

Eleocharis geniculata may be confused with *E. olivacea* in that both plants are nearly the same size and the achenes may be nearly identical except for slight color differences. *Eleocharis geniculata* typically has orbicular or ovate, obtuse spikelets while *E. olivacea* has ovate, acute spikelets. Other characteristics such as scale length and bristle appearance may be used to distinguish these species.

8. **Eleocharis intermedia** (Muhl.) Schult. in Roem. & Schult. Syst. Veg. 2:91. 1824. Fig. 109. *Scirpus intermedius* Muhl. Descr. Gram. 31. 1817, non *S. intermedius* Thuill. (1799). *Eleocharis reclinata* Kunth, Enum. 2:143. 1837.

Perennial from very short to elongated fibrous roots; culms capillary, 0.1–0.5 mm thick, 4–17 cm tall, furrowed; sheaths reddish green or straw-colored, terminating at the apex in an acute tooth; spikelets ovoid, ellipsoid, or lanceoloid, acute, 2–7 mm long, with mature achenes frequently occurring at the lowermost levels of the plant; scales ovate, occasionally subulate, obtuse to acute, 1.8–2.1 mm long,

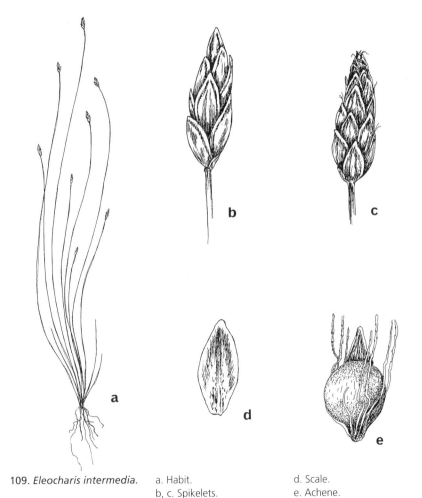

109. *Eleocharis intermedia.* a. Habit.
b, c. Spikelets.
d. Scale.
e. Achene.

dark red with a light green keel and hyaline margin or completely straw-colored, the lower sterile scale obtuse and completely encircling the culm; styles 3; achenes pyriform, 0.9–1.1 mm long (excluding the tubercle), about 0.6 mm wide, bright lemon to dark lustrous olive, faintly puncticulate; tubercle very elongated, usually with a flangelike base; bristles 6, very coarse, longer than the achenes and usually exceeding the tubercle, sometimes lacking. June–September.

Banks of rivers and streams, swamps.

IL, IN (FACW), KY, OH (FACW+).

Intermediate spikesedge.

This species, *E. acicularis*, and sometimes *E. wolfii* have capillary stems. Some of the stems of *E. wolfii* are flattened and often inrolled. *Eleocharis intermedia* differs from *E. acicularis* by having spikelets at the lowest levels of the plant.

9. **Eleocharis lanceolata** Fern. Proc. Amer. Acad. Arts 34 (19):493. 1899. Fig. 110. *Eleocharis obtusa* (Willd.) Schult. var. *lanceolata* (Fern.) Gilly, Iowa State Coll. J. Sci. 21:92. 1946.

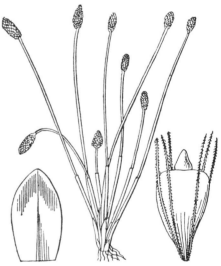

Cespitose annual; culms to 40 cm tall, 0.5–1.5 mm in diameter; sheaths red, brown, or straw-colored; spikelets lanceoloid, acute at the apex, several-flowered, 3–12 mm long; scales appressed, 1.5–2.0 mm long, ovate; style usually 2-cleft; bristles 6–7, longer than the achenes; achenes obovoid, 0.8–1.2 mm long, 0.7–0.9 mm broad, smooth, yellow to deep brown, shiny; tubercle flat-triangular, about 2/3 as wide as the achene. May–October.

Sloughs, ditches, around ponds.

KS (FACW+), MO (FACW).

Lance-spiked spikesedge.

This species has the general appearance of *E. ovata*, but differs by its narrowly lanceoloid, pointed spikelets.

110. *Eleocharis lanceolata*. Habit, scale, and achene.

10. **Eleocharis macrostachya** Britt. ex Small, Fl. S.E.U.S. 184. 1903. Fig. 111.

Perennial from creeping rhizomes; culms rigid, to 1.2 m tall, up to 2.7 mm wide, averaging 2.0 mm wide; basal sheaths truncate to slightly oblique at apex; spikelets ovoid, acute, 1.5–3.0 cm long, averaging 2.0 cm long, with usually 2 basal scales; scales ovate to lanceolate, obtuse to acute, 3.5–4.5 mm long, averaging 3.8 mm long, brown or straw-colored with a light midrib and hyaline margin; achenes obovoid, 1.5–1.6 mm long (excluding the tubercle), yellow, becoming dark brown with age; tubercle triangular-conical, about 1/6 as long as the achene. July–October.

Swamps, sloughs.

IA, IL, IN, KS, KY, MO, NE, OH (OBL).

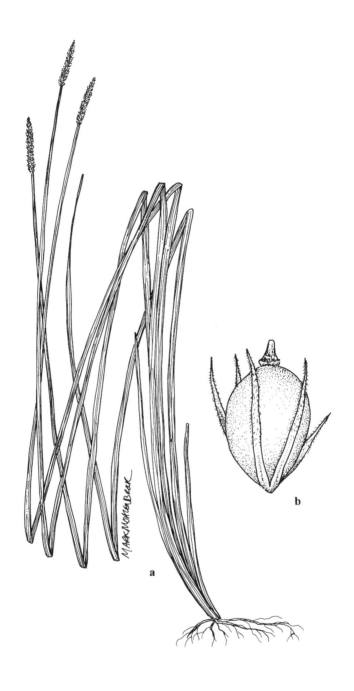

111. *Eleocharis macrostachya.*

a. Habit.

b. Achene, with bristles.

Large spikesedge.

This species has many characteristics that are similar to *E. palustris*, but at maturity this species is greater than 75 cm tall and has thicker culms.

11. **Eleocharis melanocarpa** Torr. Ann. Lyc. N.Y. 3:311. 1836. Fig. 112.

Densely tufted perennial from very short rhizomes; culms very slender, wiry, flattened, to 50 (–60) cm tall; sheaths usually brownish; spikelets narrowly ovoid, obtuse, many-flowered, 6–15 mm long; scales narrowly ovate, obtuse; bristles 0–3, much shorter than the achenes; style 3-cleft; achenes obpyramidal, trigonous, very dark brown, 0.9–1.0 mm long; tubercle flat, confluent with the achene, a little wider than the achene and slightly projecting at the edges. July–October.

Wet sand, rarely in shallow water. IN (FACW+).

Dark-fruited spikesedge.

This is primarily an Atlantic coastal plain species with disjunct localities in Indiana and Michigan. It has a very distinctive achene which is very dark brown with a flat,

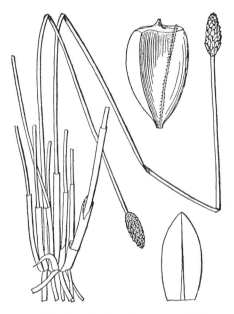

112. *Eleocharis melanocarpa.* Habit, achene, and scale.

confluent tubercle that is as wide as or slightly wider than the achene. It differs from other flat-stemmed species of *Eleocharis* by its smaller stature and absence of a twisting of some of the culms.

12. **Eleocharis microcarpa** Torr. Ann. Lyc. N.Y. 3:312. 1836. Fig. 113.

Tufted annual; culms slender, up to 30 cm tall when emergent, up to 75 cm long when submerged; sheaths brownish; spikelets ovoid to oblongoid, subacute at apex, several-flowered, 2–7 mm long; scales narrowly oblong, obtuse, often stramineous throughout; bristles several, shorter than to nearly as long as the achenes; style 3-cleft; achenes obovoid, trigonous, 0.6–0.8 mm long, greenish gray to whitish; tubercle minute. July–October.

Swamps.

IN (OBL).

Small-fruited spikesedge.

This is a species of the Atlantic coastal plain with disjunct stations in Indiana and Tennessee. *Eleocharis microcarpa* often grows submerged in water where it has elongated stems. This species is further distinguished by its very small spikelets and its small achenes with minute tubercles.

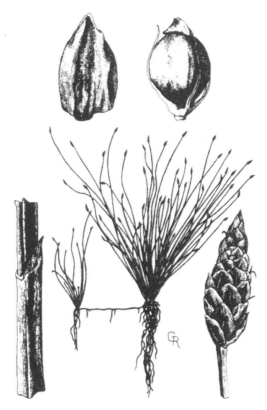

113. *Eleocharis microcarpa.* Habit, scale, achene, sheath, and spikelet.

13. **Eleocharis montevidensis** Kunth, Enum. Pl. 2:144. 1837. Fig. 114.

Perennials from thick rhizomes; culms 1.0–1.5 mm in diameter, soft, up to 50 cm tall; sheaths dark brown; spikelets ovoid to oblongoid, obtuse, many-flowered, up to 15 mm long; scales ovate to oblong, obtuse, yellow-brown with a scarious margin; bristles 4–6, shorter than to as long as the achenes; style 3-cleft; achenes obovoid, 1.0–1.2 mm long, pitted, shiny; tubercle conical, up to 1/2 as wide as the achene. July–October.

Along streams, around lakes and ponds, usually not in shallow water.

KS (FACW).

Soft-stemmed spikesedge.

Eleocharis montevidensis and *E. palustris* have similar achenes, but *E. montevidensis* has a 3-cleft style and soft stems.

14. **Eleocharis olivacea** Torr. Ann. Lyc. N.Y. 3:300. 1836. Fig. 115.

Scirpus olivaceus (Torr.) Kuntze, Rev. Gen. 2:758. 1891.

Trichophyllum olivaceum (Torr.) House, Am. Midl. Nat. 6:205. 1920.

Eleocharis flavescens (Poir.) Urban var. *olivacea* (Torr.) Gl. Phytologia 4:22. 1952.

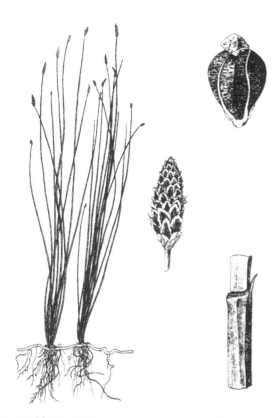

114. *Eleocharis montevidensis.* Habit, spikelet, achene, and sheath.

Perennials from thin, reddish brown or straw-colored rhizomes; culms 4–9 cm long, flat, spongy, striate or angled, 0.3–1.5 mm thick; sheaths dark red to straw-colored, hyaline at apex, the hyaline portion at times free from the culms; spikelets ovoid, acute, somewhat flattened, 3–8 mm long; scales ovate, acute to obtuse, 1.8–2.3 mm long, dark red with a light green or straw-colored keel and hyaline margin; style 2-cleft; bristles 1–6, very delicate, usually longer than the achenes, or often absent; achenes obovoid, biconvex, 1 mm long, 0.7 mm wide, dull yellow becoming lustrous olive or brown at maturity, puncticulate, with a short projection at the apex supporting the small conical tubercle. June–October.

Wet sands, rarely in shallow water.

IA, IL, IN, KY, OH (OBL).

Olive-fruited spikesedge.

This species may be recognized by its short spongy culms and its dull yellow, olive, or brown achenes that have a short projection at the apex to support the small conical tubercle. The similar appearing *E. caribaea* has orbicular to ovoid, obtuse spikelets, while *E. olivacea* has ovoid, acute spikelets.

a

b

c

115. *Eleocharis olivacea.*   a. Spikelet.   c. Scale.
                                b. Achene.

15. **Eleocharis ovata** (Roth) Roem. & Schult. Syst. Veg. 2:152. 1817.
Scirpus ovatus Roth. Catr. 1:5. 1797.
Trichophyllum ovatum (Roth) Farwell, Rep. Mich. Acad. 21:358. 1920.
Eleocharis obtusa (Willd.) Schult. var. *ovata* (Roth) Drap. & Mohlenbr. Am. Midl. Nat.
64:341. 1960.

Cespitose annuals; culms to 50 cm tall, 1–2 mm in diameter; sheaths red, brown, or straw-colored; spikelets ovoid to oblongoid, obtuse to acute, several-flowered, 2–16 mm long; scales appressed, ovate to oblong, acute; style usually 2-cleft; achenes obovoid, 0.9–1.0 (–1.3) mm long, 0.7–0.8 mm wide, smooth, yellow to deep brown; tubercle 1/2 to 2/3 as wide as the achenes; bristles shorter than to exceeding the achene. June–October.

There are three varieties in the *Eleocharis ovata* complex that intergrade into each other so that distinguishing the varieties is often difficult. This is a common species in the central Midwest. Its relatively short, thick culms are distinctive. To make matters worse, this complex used to be known as the *E. obtusa* complex, but the binomial *E. ovata* clearly predates *E. obtusa* so that the three entities are all varieties of *E. ovata*.

Since it may be useful to distinguish the three varieties, a key to the varieties of *E. ovata* is provided:

a. Tubercle more than two-thirds the width of the achene; bristles shorter than the achenes or lacking .... 5a. *E. ovata* var. *ovata*
a. Tubercle 1/2 to 2/3 the width of the achene; bristles usually longer than the achene, or lacking.
   b. Tubercle 1/4 to 1/2 the height of the achene; bristles longer than the achene, rarely absent .... 15b. *E. ovata* var. *obtusa*
   b. Tubercle up to one-fourth the height of the achene; bristles as long as or longer than the achene, or lacking .... 15c. *E. obtusa* var. *detonsa*

15a. **Eleocharis ovata** (Roth) Roem. & Schult. var. **ovata** Fig. 116.

Ditches, sloughs, sinkhole ponds, around ponds and lakes, occasionally in standing water.

IA, IL, IN, KS, KY, MO, NE, OH (OBL).

Ovate spikesedge.

Eleocharis ovata var. *ovata* is the least common of the three varieties of the species. In addition to the characteristics given in the key, the spikelets are usually a little more pointed in this variety than in the other two.

15b. **Eleocharis ovata** (Roth) Roem. & Schult. var. **obtusa** (Willd.) Kükenth. ex Skottsb. Acta Horti Gothob. 1:212. 1926. Fig. 117.
Scirpus obtusa Willd. Enum. 76. 1809.
Eleocharia obtusa (Willd.) Roem. & Schult. Syst. Veg. 2:89. 1824.

Ditches, along rivers and streams, around ponds and lakes, sinkhole ponds, sometimes in standing water.

IA, IL, IN, KS, KY, MO, NE, OH (OBL). The U.S. Fish and Wildlife Service calls this plant *E. obtusa.*

Blunt spikesedge.

This is the most common variety of *E. ovata* in the central Midwest and is one of the most common in the genus. If considered to be a distinct species, it becomes *E. obtusa,* which is the name given to it by the U.S. Fish and Wildlife Service.

15c. **Eleocharis ovata** (Roth) Roem. & Schult. var. **detonsa** (Gray) Mohlenbr. Illus. Fl. Ill. Sedges, ed. 2, 195. 2001. Fig. 118.
Eleocharis engelmanni Steud. Syn. Cyp. 79. 1855.
Eleocharis engelmanni Steud. var. *detonsa* Gray in Patterson, List Pl. Vic. Oquawka 17. 1874.
Eleocharis ovata (Roth) Roem. & Schult. var. *engelmanni* (Steud.) Britt. Journ. N.Y. Micr. Soc. 5:103. 1889.

Ditches, margins of ponds and lakes, swamps, occasionally in standing water.

IA, IL, IN, KS, KY, MO, NE, OH. The U.S. Fish and Wildlife Service calls this plant *E. engelmannii.* It is FACW in IA, IL, IN, KS, MO, NE, and FACW+ in KY, OH.

Engelmann's spikesedge.

If this plant is considered to be a distinct species, it would be *E. engelmannii.*

16. **Eleocharis palustris** (L.) Roem. & Schult. Syst. Veg. 2:151. 1817. Fig. 119.
Scirpus palustris L. Sp. Pl. 1:47. 1753.
Eleocharis smallii Britt. Torreya 3:23. 1903.

Culms 60 cm tall, 0.6–1.8 mm in diameter, averaging 1.1 mm, often with a reddish base; spikelets ovoid to ellipsoid to lanceoloid, acute to acuminate with a cuspidate tip, 7–20 mm long, averaging 12 mm, with 2 or rarely 3 basal scales; scales ovate to lanceolate, obtuse to acute to acuminate, 2.7–4.0 mm long, averaging 3 mm, brown or straw-colored with a light midrib and hyaline margins; achenes obovoid, 2.3–2.6 mm long (excluding the tubercle), yellow, becoming dark brown

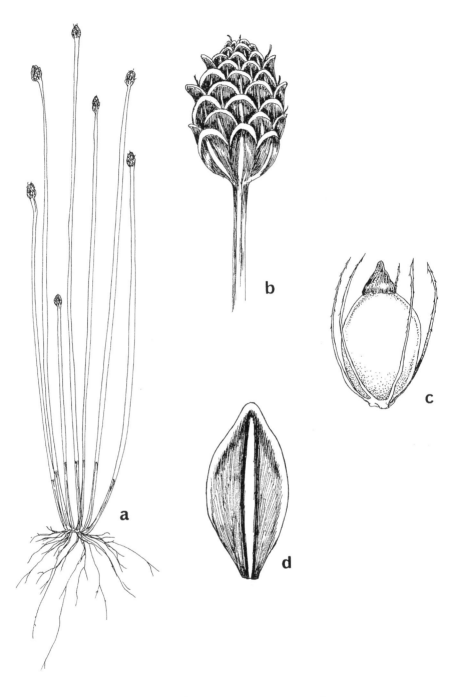

116. *Eleocharis ovata*
var. *ovata*.

a. Habit.
b. Spikelet.

c. Achene.
d. Scale.

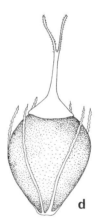

117. *Eleocharis ovata*
var. *obtusa.*

a. Habit.
b. Spikelet.

c. Scale.
d. Achene.

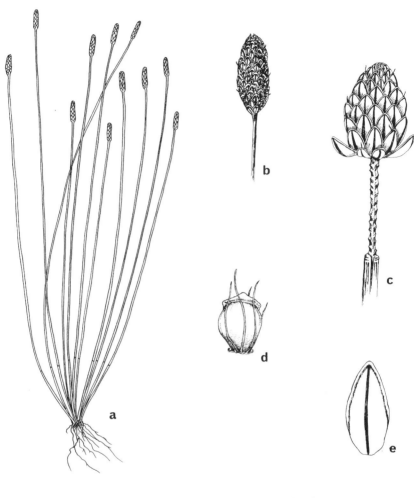

118. *Eleocharis ovata* var. *detonsa*.

a. Habit.
b, c. Spikelets.
d. Achene.
e. Scale.

with age; tubercle triangular-conic, up to 1/2 as long as the achene; bristles 3–6, longer than the achene. June–October.

Ditches, around ponds and lakes, along rivers and streams, sloughs, swampy forests, marshes, sometimes in shallow water.

IA, IL, IN, KS, KY, MO, NE, OH (OBL).

Marsh spikesedge.

This is often the most common species of *Eleocharis* in the central Midwest. The culms up to 60 cm tall and averaging 1.1 mm in diameter are distinctive. *Eleocharis smallii* may indeed be a distinct species, usually with the bases of the culms reddish.

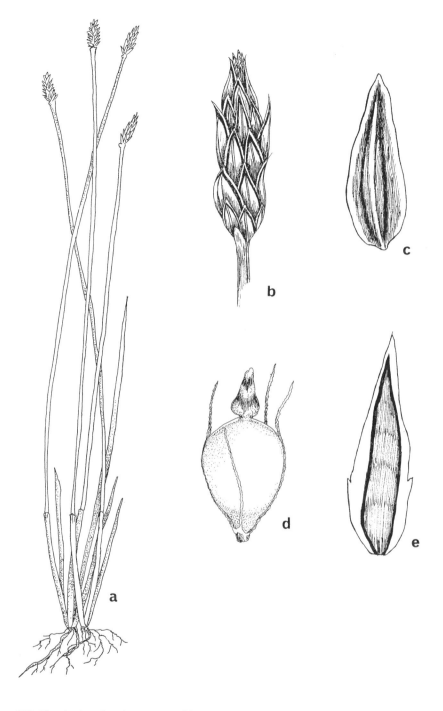

119. *Eleocharis palustris.* a. Habit.
b. Spikelet.

c, e. Scales.
d. Achene.

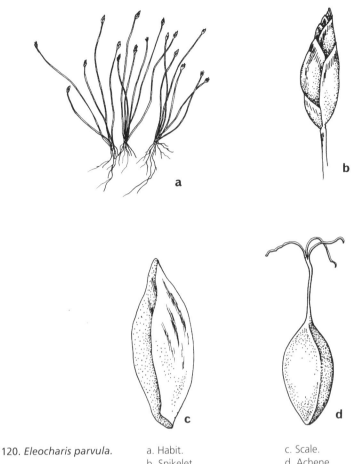

120. *Eleocharis parvula.*

a. Habit.
b. Spikelet.

c. Scale.
d. Achene.

17. **Eleocharis parvula** (Roem. & Schult.) Link ex Bluff, Nees, & Schauer in Bluff and Fingerhuth, Comp. Fl. Germ., 3d. 2, 1:93. 1836. Fig. 120.
Scirpus nanus Spreng. Pug. 1:4. 1813, non Poir. (1804).
Scirpus parvulus Roem. & Schult. Syst. 2:124. 1817.
Eleocharis pygmaea var. *anachaeta* Torr. Ann. Lyc. N.Y. 3:441. 1836.
Eleocharis parvula (Roem. & Schult.) Link var. *anachaeta* (Torr.) Svenson, Rhodora 35: 386. 1933.

Densely tufted perennial from very slender stolons, the stolons often bearing brownish tubers up to 6 mm long; culms filiform, stiff, to 7 cm tall; spikelets ovoid, somewhat flattened, acute, 2- to 9-flowered, 2–4 mm long; scales ovate, acute, 1.5–2.5 mm long, pale brown to green; stamens 3; style 3-cleft; achenes trigonous, oblongoid, 1.0–1.5 mm long, pale brown, the style scarcely differentiated from the body of the achene as a minutely triangular tubercle; bristles shorter than or barely exceeding the achenes, or absent. July–October.

Usually around ponds, rarely in standing water.

IA, IL, KS, MO, OH (OBL).

Small spikesedge.

This species has the dwarf stature of *E. acicularis,* but the culms are filiform and stiff rather than capillary and flaccid.

Typical plants lack bristles that subtend the achenes. If bristles are present, those plants may be called var. *anachaeta* (Torr.) Svenson.

18. **Eleocharis pauciflora** (Lightf.) Link, Hort. Berol. 1:284. 1827. Fig. 121.
Scirpus quinqueflora F. X. Hartman, Prin. Linn. Inst. Bot. Crantzii, ed. 2, 85. 1767, misapplied.
Scirpus pauciflorus Lightf. Fl. Scot. 1078. 1777.
Eleocharis pauciflora (Lightf.) Link var. *fernaldii* Svenson, Rhodora 36:380. 1934.
Eleocharis quinqueflora (F. X. Hartman) Schwarz, Mitt. Thuring. Bot. Ges. 1:89. 1949.

Perennials from slender stoloniferous rhizomes, often bearing light brown scaly tubers about 2.2 mm long; culms 76–21 cm tall, 0.2–0.4 mm in diameter, red, brown, or straw-colored at the base, usually angled; spikelets ellipsoid to ovoid, acute, 2- to 7-flowered, 3–7 mm long; scales lanceolate, acute, 4–5 mm long, reddish brown with a hyaline margin, spreading at maturity; stamens 3; style 3-cleft; achenes trigonous, obovoid or fusiform, 2.1–2.5 mm long, 0.9–1.0 mm wide, finely reticulated, dull yellow, turning light olive at maturity, the style often barely differentiated from the body of the achenes, the style often black-tipped; bristles yellow, shorter or longer than the achenes, or absent. July–October.

Around ponds and lakes, sometimes in shallow water.

IA, IL, IN, NE, OH (OBL). The U.S. Fish and Wildlife Service calls this species *E. quinqueflora.*

Few-flowered spikesedge.

Eleocharis pauciflora is distinguished readily by its conical tubercles, which are barely differentiated from the bodies of the dull yellow or olive achenes. *Eleocharis rostellata* has similar tubercles, but the culms of *E. rostellata* are longer and thicker.

19. **Eleocharis quadrangulata** (Michx.) Roem. & Schult. Syst. Veg. 2:155. 1817. Fig. 122.
Scirpus quadrangulatus Michx. Fl. Bor. Am. 1:30. 1803.
Eleocharis quadrangulata (Michx.) Roem. & Schult. var. *crassior* Fern. Rhodora 37:393. 1935.

Perennial from rather thick, brown or red, coarse, fibrous roots; culms 4-sided, sharply angled, 6–14 dm high, 2–5 mm thick; sheaths dark red or brown, often with a loose, leaflike tip; spikelets 2–6 cm long, acute, as thick as or slightly thicker than the culms; scales elliptic, obtuse, straw-colored, with a hyaline margin, 6 mm long, 3.5 mm wide; stamens 3; style 2- or 3-cleft; achenes biconvex, obovoid, tapering below the base of the tubercle, greenish yellow or tan, becoming dark lustrous brown upon maturity, 2.1–2.5 mm long, 1.5–1.6 mm wide; tubercle conical, about 1/2 the length of the achenes; bristles 6–8, slightly longer than the achene. July–October.

Shallow water in ponds and lakes.

IL, IN, KS, KY, MO, OH (OBL).

Square-stemmed spikesedge.

This is the only member of the Cyperaceae with a four-sided culm.

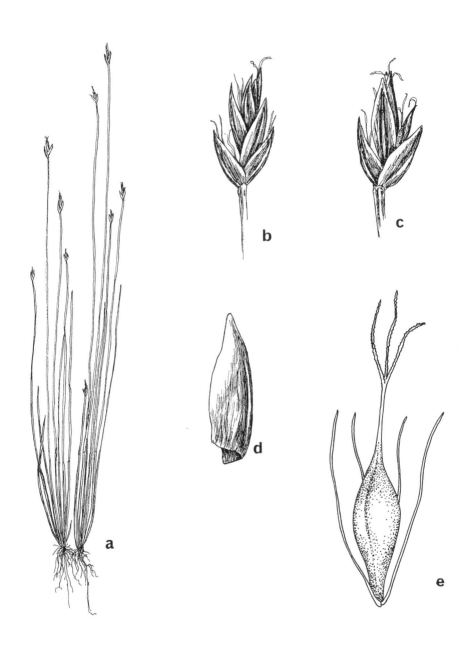

121. *Eleocharis pauciflora.*

a. Habit.
b, c. Spikelets.
d. Scale.
e. Achene.

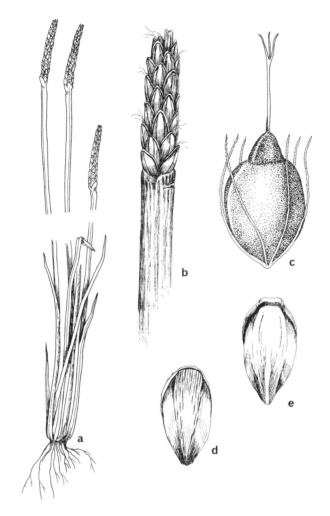

122. *Eleocharis quadrangulata.*

a. Habit.
b. Spikelet.

c. Achene.
d, e. Scales.

20. **Eleocharis robbinsii** Oakes, Hovey's Mag. Bot. 7:178. 1841. Fig. 123.

Tufted perennial from slender rhizomes; fertile culms up to 70 cm tall, 1–2 mm in diameter, triangular; sterile culms, when present, floating, capillary; sheaths brown; spikelets lanceoloid, acute to acuminate at the apex, up to 2.0 (–2.5) cm long, 4- to 8-flowered; scales narrowly ovate, obtuse, with scarious margins; style 2-cleft; achenes obovoid, lenticular, 2.0–2.5 mm long, brown, constricted above into a distinct neck; tubercle flat, acuminate, up to 1/2 as long as the achenes; bristles toothed, longer than the achenes. July–October.

Shallow water.

IN (OBL).

Robbins' spikesedge.

This mostly Atlantic coastal plain species has inland locations in Indiana, Michigan, and Wisconsin.

Of the three species of *Eleocharis* in the central midwest with spikelets as thick as or thicker than the culms, this is the only one with triangular, non-septate culms. The narrowly lanceoloid spikelets, the achenes with a constricted neck, and the acuminate tubercles are also distinctive.

21. **Eleocharis rostellata** (Torr.) Torr. Fl. N.Y. 2:347. 1843. Fig. 124. *Scirpus rostellatus* Torr. Ann. Lyc. N.Y. 3:318. 1836.

Plants densely tufted; culms terete, 0.1–1.5 m tall, wiry, occasionally rooting at the tip;

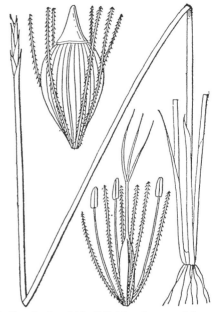

123. *Eleocharis robbinsii.* Habit, achene, and flower.

spikelets nearly ovoid, subacute, several-flowered, 5–20 mm long; scales elliptic, obtuse, 3.5–5.0 mm long, brownish, coriaceous; style usually 3-cleft; achenes trigonous to obovoid, narrowed to the tubercle, 1.0–2.2 mm long; tubercle confluent, pyramidal, 1–2 mm long; bristles firm, as long as the achene and tubercle together. July–October.

Marshy soils, sedge meadows, often calcareous, uncommon in shallow water. IL, IN, KS, NE, OH (OBL).

Slender spikesedge.

The pyramidal tubercle, which is confluent with the achene, is similar to that found in *E. pauciflora*, but this latter species has angular stems and acute scales.

22. **Eleocharis verrucosa** (Svenson) Harms, Am. Journ. Bot. 59 (5):486. 1972. Fig. 125. *Eleocharis capitata* (L.) Blake var. *verrucosa* Svenson, Rhodora 34:202. 1932. *Eleocharis tenuis* (Willd.) Schult. var. *verrucosa* (Svenson) Svenson, Rhodora 41:66. 1939.

Perennials from long, slender rhizomes 1–4 mm thick, purple or brown; culms 5-angled, capillary, 0.6–7.0 dm long, 0.3–0.6 mm thick; sheaths reddish purple or straw-colored, closely adherent to the culms; spikelets ovoid to lanceoloid, obtuse to acute, 2–10 mm long; scales ovate, obtuse to acute, 1.0–2.5 mm long, dark red or purple with a light margin; styles 3-cleft; achenes trigonous, obovoid, 0.6–0.9 mm long, conspicuously warty, dark olive to yellow; tubercle conical, depressed or depressed with an apiculate center, all types sometimes being found on the same plant; bristles absent. May–September.

Around ponds and lakes, ditches, sloughs, sinkhole ponds, wet depressions in forests, uncommon in shallow water.

124. *Eleocharis rostellata.*
a. Habit.
b. Scale.

c. Achene.

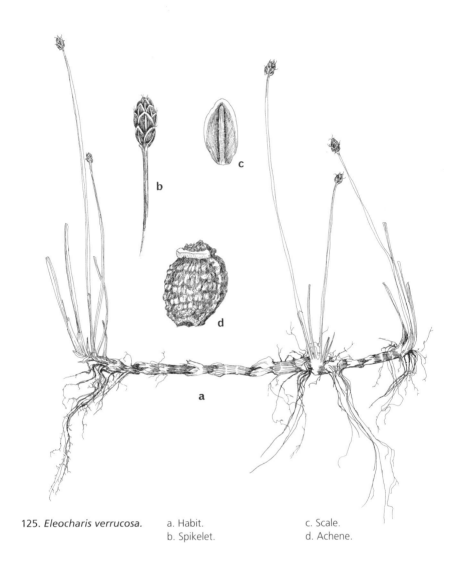

125. *Eleocharis verrucosa.* a. Habit.
b. Spikelet.
c. Scale.
d. Achene.

IA, IL, IN, KS, KY, MO, NE, OH. The U.S. Fish and Wildlife Service does not recognize this species, considering it to be the same as *Eleocharis tenuis. Eleocharis tenuis,* however, is a different species in my opinion. *Eleocharis tenuis* is considered FACW by the Fish and Wildlife Service in IA, IL, IN, KS, MO, NE, and FACW+ in KY and OH.

Warty spikesedge.

The 5-angled culms, dark red or purple scales, and olivaceous or yellow, warty achenes distinguish *E. verrucosa.*

23. **Eleocharis wolfii** (Gray) Patterson, Cat. Pl. Ill. 46. 1876. Fig. 126.
Scirpus wolfii Gray, Proc. Am. Acad. 10:77. 1874.

Perennials from light brown, stoloniferous rhizomes, commonly in clumps; culms very flat, furrowed to nearly striate, often inrolled, sometimes twisted, 8–38 cm

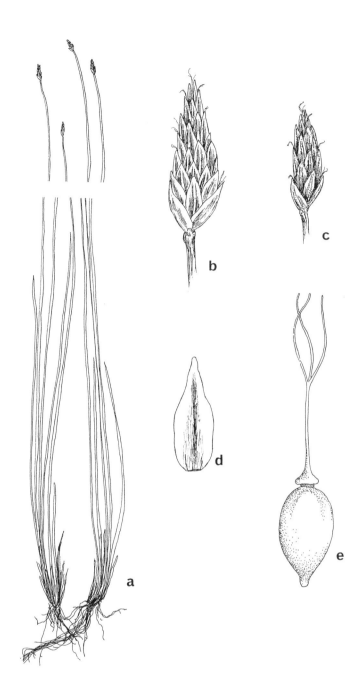

126. *Eleocharis wolfii.*   a. Habit.
                            b, c. Spikelets.

d. Scale.
e. Achene.

long, 0.3–1.5 mm wide; sheaths dark red, brown, or straw-colored, terminating at the apex as a thin, hyaline tooth that may be split and free from the culm; spikelets ovoid, ellipsoid, or lanceoloid, acute, 4–9 mm long; scales ovate to lanceolate, acute, 2.5–3.0 mm long, dark red with a broad light green or brown midvein and hyaline margins to straw-colored with hyaline margins, to nearly hyaline throughout, the 2 lower basal scales usually larger; style 3-cleft; achenes obovoid to ellipsoid, 0.7–1.0 mm long, 0.4–0.5 mm wide, light brown or nearly white, with 9–19 longitudinal ridges and numerous transverse trabeculae; tubercle conical, depressed, or depressed with an apiculate center, a constriction usually separating the tubercle from the achene; bristles absent. May–July.

Around ponds, ditches, wet depressions in forests, uncommon in shallow water. IA, IL, IN, MO, OH (OBL).

Wolf's spikesedge.

This rather rare spikesedge often forms mats similar to *E. acicularis.* The stems of *E. wolfii,* however, are flattened and often twisted.

24. **Eleocharis xyridiformis** Fern. & Brackett, Rhodora 31:76–77. 1929. Fig. 127.

Perennial from stout, black rhizomes; culms flat, stiff, often twisted, up to 50 cm tall; sheaths reddish to purple; spikelets narrowly lanceoloid, acute, many-flowered, up to 2 cm long; scales elliptic to ovate, obtuse, with scarious margins, the basal scale of each spikelets orbicular; style 2-cleft; achenes obovoid, lenticular, 1.5–1.8 mm long, 1.3–1.5 mm wide, stramineous to dark brown; tubercle deltoid, about 0.5 mm long; bristles usually 4–6, almost as long as the achenes, rarely absent. June–October.

Ditches, around lakes and ponds, swamps, marshes.

KS, NE. The U.S. Fish and Wildlife Service does not distinguish this species from *E. smallii,* which it lists as OBL.

Xyrislike spikesedge.

This species is readily distinguished by its flat, twisted culms and 2-cleft styles.

7. **Eriophorum** L.—Cotton Sedge

Slender to robust tufted perennials, with terete to angular culms; leaves usually flat at the base, becoming triangular-channeled to nearly folded at the subulate tip, sometimes channeled throughout; inflorescence composed of solitary or clustered, umbellate spikelets subtended by 1–5 erect or spreading involucral bracts; spikelets several-flowered, with the scales elongating during anthesis; bristles numerous, white or cream, becoming greatly elongated in fruit; stamens 1–3; styles 3-cleft; ovary superior, 1-locular; achenes sharply trigonous.

There are 20 species in this genus, all found in the northern hemisphere. Some botanists have proposed merging *Eriophorum* with *Scirpus,* but I prefer to retain *Eriophorum* as a separate genus because of the "cotton heads" produced during fruiting.

1. Spikelet 1 per culm; lowest scales sterile; bracts absent .... 3. *E. spissum*
1. Spikelets at least 2 per culm; lowest scales fertile; bracts 1–3.
  2. Leaves up to 1.5 mm wide, channeled throughout; involucral bract 1.
    3. Culms glabrous throughout; sheath of leaf longer than blade; achenes 1.5–2.0 mm long .... 2. *E. gracile*

127. *Eleocharis xyridiformis.* Habit and four spikelets.

3. Culms scabrous in the upper half; sheath of leaf shorter than blade; achenes 2.5–
   3.1 mm long .... 4. *E. tenellum*
2. Leaves 1.5–8.0 mm wide, channeled only above the middle; involucral bracts 2–5.
   4. Stamens 3; scales one-nerved.
      5. Peduncles glabrous; anthers 2.5–5.0 mm long .... 1. *E. angustifolium*
      5. Peduncles puberulent; anthers 1.0–1.3 mm long .... 6. *E. viridi-carinatum*
   4. Stamen 1; scales several-nerved .... 5. *E. virginicum*

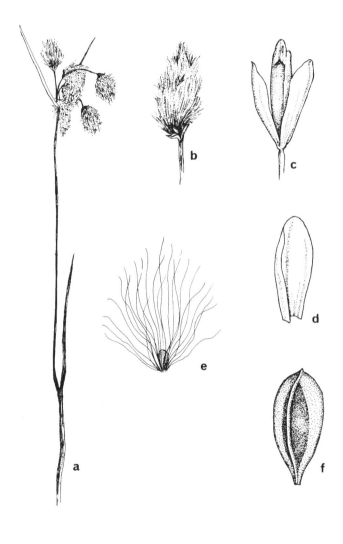

128. *Eriophorum angustifolium.*
a. Habit.

b. Spikelet.
c. Spikelet with bristles removed.

d. Scale.
e. Achene.
f. Achene with bristles removed.

1. **Eriophorum angustifolium** Honck. Veregz. Gew. Teutschl. 153. 1782. Fig. 128.
Eriophorum polystachion L. Sp. Pl. 1:52. 1753.
Eriophorum angustifolium Honck. var. *majus* Schultz, Fl. Starg. Suppl. 5. 1819.
Eriophorum polystachion L. var. *elatius* Babington, Man. Brit. Bot. 333. 1843.

Rather robust, tufted perennial; culms angled, glabrous, to 80 cm tall; leaves channeled in the upper half, flattened below, 3.5–7.5 mm wide, with scabrous margins; involucral bracts 2–3, flattened below, more or less channeled above, purple near the base; spikelets 2–10, on glabrous peduncles to 5.5 cm long, the spikelets ovoid, 1–2 cm long in flower, 2.5–5.0 cm long in fruit; scales lead-colored to castaneous, acute, 1-nerved, the nerve not reaching the tip of the scale; anthers 2.5–5.0 mm long; achenes 2.5–3.5 mm long, apiculate; bristles numerous, white. June–August.

Bogs, fens, marshes, swamps, ditches.

IA, IL, IN, NE (OBL).

Thin-scaled cotton sedge.

The combination of 2–3 bracts and 1-nerved scales in which the nerve fails to reach to tip of the scale distinguishes this species. A few botanists call this species *E. polystachion*, a European species which, in my opinion, is a different species.

2. **Eriophorum gracile** Koch ex Roth, Cat. Bot. 2:259. 1800. Fig. 129.

Perennial from slender rhizomes; culms slender, terete, glabrous, to 60 cm tall; leaves channeled throughout, to 4 cm long, to 1.5 mm broad, rounded at the apex, the uppermost sheath longer than the blade; involucral bracts 2–5, erect, channeled throughout, 1–2 cm long, dark gray to black at base; spikelets 2–5 (–6) on slender, puberulenet peduncles up to 3 cm long, the spikelets ellipsoid to ovoid, 7–10 mm long in flower, 12–18 mm long in fruit; scales lead-colored or blackish, obtuse or subacute; anthers 1.2–2.0 mm long; achenes 1.5–2.0 mm long; bristles numerous, white. April–July.

Bogs, swamps, shores.

IA, IL, IN, NE, OH (OBL).

Slender cotton sedge.

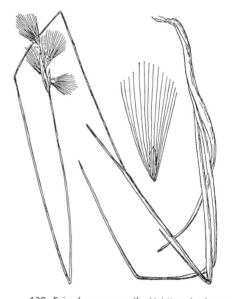

129. *Eriophorum gracile.* Habit and achene.

This species has 2 or more bracts, completely glabrous culms, and achenes 1.5–2.0 mm long.

3. **Eriophorum spissum** Fern. Rohdora 27:208–210. 1925. Fig. 130.
Eriophorum vaginatum L. ssp. *spissum* (Fern.) Hulten, Acta Univ. Lund.
38 (1):286. 1942.

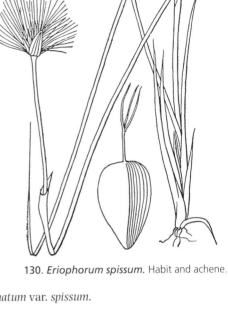

Tufted perennial forming large
clumps; culms to 70 cm tall;
leaves numerous, nearly all near
the base of the plant, filiform,
0.8–1.2 mm wide; sheaths at
middle of stem bladeless; involucral bracts absent; spikelet one,
1.0–1.5 cm long in flower, 2–3 cm
long in fruit, with 10–15 sterile
basal scales; scales narrowly
ovate, long-acuminate, blackish
with pale margins; bristles
numerous, usually white; anthers
2–3 mm long; achenes obovoid,
2.5–3.5 mm long, minutely
apiculate. April–July.

Bogs, conifer swamps.

130. *Eriophorum spissum*. Habit and achene.

IN (OBL). The U.S. Fish and
Wildlife Service calls this plant *E. vaginatum* var. *spissum*.

Tussock cotton sedge.

This is the only *Eriophorum* in the central Midwest with a solitary spikelet per
culm and no involucral bracts.

4. **Eriophorum tenellum** Nutt. Gen. Am. 2 Add.
1818. Fig. 131.
Eriophorum gracile Koch var.
paucinervum Engelm. Am. Journ. Sci.
45:105. 1846.
Eriophorum paucinervum (Engelm.)
A. A. Eaton, Bull. Torrey Club
25:341. 1898.

Perennial from slender roots; culms
slender, scabrous (at least in the upper
half), to 80 cm tall; leaves channeled
throughout, to 15 cm long, up to 1.5 mm
broad, subulate, the uppermost sheath
shorter than the blade; involucral
bract 1, channeled throughout, to
6 cm long, usually reddish brown
near its base; spikelets 2–7, on
scabrous peduncles to 2 cm long, the
spikelets 1.2–1.7 cm long in flower, 2–

131. *Eriophorum tenellum*. Habit and achene.

3 cm long in fruit; scales greenish or reddish brown, more or less obtuse; anthers 1–2 mm long; achenes 2.5–3.1 mm long, apiculate; bristles numerous, white. June–September.

Bogs, conifer swamps.

IL (OBL).

Conifer cotton sedge.

The scabrous upper part of the culm is distinctive for this species.

5. **Eriophorum virginicum** L. Sp. Pl. 1:52. 1753. Fig. 132.
Eriophorum virginicum L. var. *album* Gray, Man. Bot., ed. 5, 566. 1867.
Eriophorum virginicum L. f. *album* (Gray) Wieg. Rhodora 26:2. 1924.

Perennial from slender roots; culms slender, glabrous, to 1 m tall; leaves channeled only at the tip, to 4 mm wide; involucral bracts 2–3, spreading, channeled at the tip, to 10 cm long; spikelets up to 20, crowded in a glomerule, ellipsoid, 6–10 mm long in flower, up to 20 mm long in fruit; scales tawny or stramineous or green with brownish margins, acute, several-nerved; stamen 1; anthers 1.2–1.5 mm long; achenes 3–4 mm long; bristles numerous, tawny or white. May–August.

Bogs, conifer swamps, marshes.

IA, IL, IN, OH (OBL).

Tawny cotton sedge.

The tawny scales, 2–3 involucral bracts, and several-nerved scales distinguish *E. virginicum*. White bristles are found in var. *album*.

6. **Eriophorum viridi-carinatum** (Engelm.) Fern. Rhodora 7:89. 1905. Fig. 133.
Eriophorum latifolium L. var. *viridi-carinatum* Engelm. Am. Journ. Sci. 46:103. 1843.
Eriophorum polystachion L. var. *latifolium* (L.) Gray, Man. Bot., ed. 5, 529. 1867.

Tufted perennial from fibrous roots; culms slender, triangular, glabrous, to 75 cm tall; leaves channeled at the tip, to 6 mm wide; involucral bracts 2–4, channeled only at the tip, green or pale brown at the base; spikelets (2–) 3–25, on puberulent peduncles to 6 cm long, the spikelets elliptic-ovoid, 6–10 mm long in flower, 15–30 mm long in fruit; scales green or occasionally lead-colored, sometimes mucronulate, 1-nerved, the nerve reaching the tip of the scale; anthers 1.0–1.3 mm long; achenes 2.0–3.5 mm long, apiculate; bristles numerous, cream-colored.

Bogs, conifer swamps, marshes.

IL, IN, OH (OBL).

Dark-scaled cotton sedge.

This species has cream-colored bristles, 2 or more involucral bracts, and 1-nerved scales with the nerve reaching the tip of the scales.

8. **Fimbristylis** Vahl—Fimbry

Tufted annuals or perennials; leaves 1–several, very narrow; inflorescence cymose, or all spikelets sessile; spikelets ovoid-cylindric, several-flowered; scales imbricate in several ranks; flowers perfect; perianth none; stamens 1–3; styles 2- or 3-cleft, the base not persisting as a tubercle; ovary superior, 1-locular; achenes lenticular or trigonous, smooth, reticulate or ribbed.

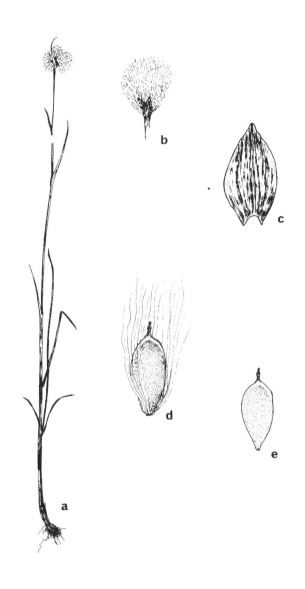

132. *Eriophorum virginicum.*
a. Habit.

b. Spikelet.
c. Scale.

d. Achene.
e. Achene with bristles removed.

133. *Eriophorum*
viridi-carinatum.

a. Habit.
b. Spikelet.

c. Scale.
d. Achene.

This genus is more prevalent in tropical and subtropical regions of the world where there are approximately 250 species. Although most species are found in wetlands, they only occasionally are found in standing water.

1. Style branches 3; achenes trigonous.
    2. Spikelets lanceoloid to narrowly ellipsoid, acute at tip, 3–7 mm long .... 2. *F. autumnalis*
    2. Spikelets ovoid to oblongoid, obtuse at tip, 2–4 mm long .... 3. *F. littoralis*
1. Styles branches 2; achenes lenticular.
    3. Spikelets in a close head; achenes 0.4–0.5 mm long .... 4. *F. vahlii*
    3. Spikelets on elongated rays, cymose; achenes 1.0–1.3 mm long .... 1. *F. annua*

1. **Fimbristylis annua** (All.) Roem. & Schult. Syst. 2:95. 1817. Fig. 134.
Scirpus annuus All. Fl. Ped. 2:277. 1785.
Scirpus baldwinianus Schult. in Roem. & Schult. Syst. Veg. Mant. 2:85. 1824.
Fimbristylis baldwiniana (Schult.) Torr. Ann. Lyc. N.Y. 3:344. 1836.

Cespitose annual; culms to 40 cm tall; leaves linear, flat, to 2 mm wide, ciliate, shorter than the culms; inflorescence umbelliform to cymose, the rays simple to branched; spikelets obovoid to obovoid-oblong, 3–10 mm long; scales ovate, obtuse to acute, rarely mucronulate, glabrous, brown with a pale, conspicuous midvein; achenes obovoid, lenticular, pale gray to stramineous, 1.0–1.3 mm long, with horizontal and transverse ribs. June–October.

Around lakes and ponds, along rivers and streams, rarely in shallow water.

IL, KS, MO (FACW-).

Annual fimbry.

The combination of 3 style branches, glabrous scales, and achenes at least 1 mm long distinguishes this rather rare species.

2. **Fimbristylis autumnalis** (L.) Roem. & Schult. Syst. 2:97. 1817. Fig. 135.
Scirpus autumnalis L. Mant. 2:180. 1771.
Scirpus mucronulatus Michx. Fl. Bor. Amer. 1:31. 1803.
Trichelostylis borealis Wood, Bot. & Fl. ed. 1871, 364. 1871.
Fimbristylis autumnalis (L.) Roem. & Schult. var. *mucronulata* (Michx.) Fern. Rhodora 37:398. 1935.

Cespitose annual; culms to 20 cm tall, slender, glabrous; leaves linear, flat, 2.5 (–4.0) mm wide, glabrous; inflorescence umbelliform to cymose, with simple or branched rays; bracts 2–3, shorter than to longer than the inflorescence; spikelets ovoid to ovoid-cylindric, 3–7 (–10) mm long, sessile or pedunculate; scales ovate to ovate-lanceolate, usually keeled, acute, mucronate, brown; styles 3-cleft; achenes obovoid, trigonous, 0.5–1.0 mm long, gray to pale brown, faintly reticulate to verrucose. June–October.

Around ponds and lakes, along rivers and streams, in moist depressions, on sandbars.

IA, IL, IN, KY, MO, OH (FACW+), KS (OBL).

Common fimbry.

134. *Fimbristylis annua.*

a. Habit.
b, c. Spikelets.
d. Achene.
e, f. Scales.

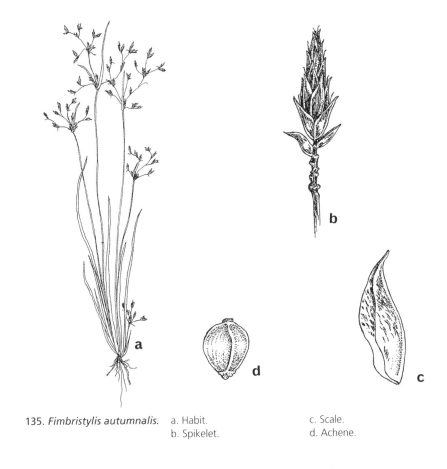

135. *Fimbristylis autumnalis.*   a. Habit.   c. Scale.
                                   b. Spikelet.   d. Achene.

This is the most widespread and abundant species of *Fimbristylis* in the central Midwest. It is readily distinguished by its 3-cleft styles, trigonous achenes, and spikelets 3–7 mm long.

3. **Fimbristylis littoralis** Gaudich, Voy. Uranie 413. 1826. Fig. 136.

Cespitose annual; culms to 35 (−50) cm tall, slender, glabrous; leaves linear, flat, 2.5–3.5 mm wide; inflorescence umbelliform, subtended by 2–3 bracts shorter than the inflorescence; spiklelets ovoid to oblongoid, obtuse, 2–4 mm long, pedunculate; scales broadly ovate, obtuse, 0.8–1.2 mm long, brown; styles 3-cleft; achenes obovoid, slightly trigonous, 0.9–1.1 mm long, reticulate, verrucose. June–October.

Wet soil, rarely in standing water.

MO (NI). Not listed from MO by the U.S. Fish and Wildlife Service.

Southern fimbry.

This southern species has been found once in the central Midwest. It is very similar in appearance to the subtropical *F. miliacea* (L.) Vahl. It also resembles the common *F. autumnalis* but differs by its smaller, ovoid, obtuse spikelets.

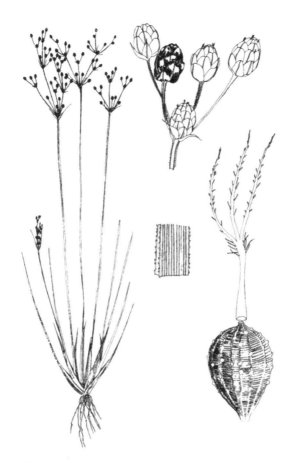

136. *Fimbristylis littoralis.* Habit, spikelet, section of stem, and achene.

4. **Fimbristylis vahlii** (Lam.) Link, Hort. Berol. 1:287. 1827. Fig. 137.
Scirpus vahlii Lam. Tab. Encyc. 1:139. 1791.

   Cespitose annual; culms capillary, rarely over 10 cm tall, glabrous; leaves filiform, more or less scabrous, as long as to barely longer than the culms; inflorescence glomerulate, with 2–4 filiform, glabrous bracts much exceeding the spikelets; spikelets 3–8, more or less obtusely cylindric, 4–8 mm long, many-flowered; scales lanceolate to oblong-lanceolate, acute to acuminate, glabrous, with a prominent midrib; styles 2-cleft; stamen 1; achenes lenticular, stramineous, shiny, 0.4–0.5 mm long, transversely reticulate. June–October.
   Sloughs, along rivers, around ponds, rarely in shallow water.
   IL, MO (OBL), KS (FACW), KY (NI).
   Dwarf fimbry.
   This tiny fimbry often forms dense mats. It is distinguished by its headlike clusters of spikelets and its very tiny achenes that are 0.4–0.5 mm long.

137. *Fimbristylis vahlii.*
a. Habit.
b. Inflorescence.
c, e. Spikelets.
d. Scale.
f. Achene.

Superficially *Fimbristylis vahlii* resembles *Isolepis carinatus* and the three species of *Hemicarpha* because of its small stature and filiform or capillary leaves and culms. The following table summarizes the differences:

Fimbristylis vahlii	*Isolepis carinatus*	*Hemicarpha* spp.
spikelets 3–8	spikelet 1	spikelets 1–3
spikelets 4–8 mm long	spikelets 3–6 mm long	spikelets 2–5 mm long
achene 0.4–0.5 mm long	achene 1.5 mm long	achene 0.5–0.8 mm long
bracts 2–4	bract 1	bracts 1–3

9. **Fuirena** Rottb.—Umbrella Sedge; Bur Sedge

Perennial herbs, often with thickened rhizomes; culms usually leafy (sometimes reduced to sheaths), more or less 3-angular; inflorescence in terminial and often axillary heads, umbels, or irregular clusters; spikelets usually several, occasionally solitary, usually many-flowered; scales spiralny arranged, awned or mucronate; bristles usually 6, the inner 3 usually scalelike and thickened, the outer 3 often much elongated; stamens 3; style 3-cleft; achenes trigonous, usually with a beak about as long as or longer than the achene.

This genus is recognized by having 3 scalelike thickened bristles that alternate with 3 elongated slender bristles. The genus consists of about 30 subtropical and tropical species.

1. Leaves reduced to bladeless sheaths .... 2. *F. scirpoidea*
1. Leaves with blades.
    2. Rhizomatous perennial .... 3. *F. simplex*
    2. Tufted annual .... 1. *F. pumila*

1. **Fuirena pumila** Michx. Fl. Bor. Am. 1:38. 1803. Fig. 138.

Cespitose annual; culms to 50 (–60) cm tall; sheaths more or less hispid; leaves linear, 3–5 mm wide, pubescent or at least scabrous; spikelets usually several in 1–3 clusters, 7–15 mm long; scales puberulent, 2.5–3.5 mm long, with a recurved awn; scaly bristles long-clawed at base, long-acuminate at apex; achenes sharply trigonous, stipitate. June–October.

Shores of ponds and lakes, rarely in shallow water.

IN (OBL).

Small umbrella sedge.

The spikelets appear bristly because of the long, recurved awns of the scales. *Fimbristylis simplex* is very similar but usually possesses slender rhizomes.

138. *Fuirena pumila.* Habit, achene, and spikelet.

2. **Fuirena scirpoidea** Michx. Fl. Bor. Am. 1:38. 1803. Fig. 139.

Perennial from slender, woody rhizomes; culms to 40 cm tall, glabrous to slightly pubescent; leaves reduced to bladeless, brownish sheaths, the sheaths loose and up to 2.5 cm long; spikelet solitary, terminal, to 12 mm long; scales puberulent, mucronate, 2–3 mm long; bristles reduced to ovate, acute scales 0.5–1.0 mm long, each borne on a stipe 0.5–1.0 mm long; achenes trigonous, dark brown to black, 1.5–2.5 mm long, tapering to a hispidulous beak.

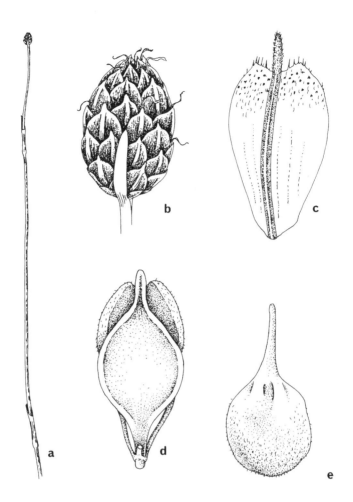

139. *Fuirena scirpoidea.*
a. Habit.

b. Spikelet.
c. Scale.

d. Achene with scales.
e. Achene.

Edge of pond.

IL. This species is not listed from the north-central region by the U.S. Fish and Wildlife Service, but it is OBL elsewhere.

Leafless umbrella sedge.

The Illinois station is hundreds of miles north of the nearest location for this species. It is readily distinguished by its bladeless sheaths. It most nearly resembles an *Eleocharis* because of its solitary spikelet and absence of leaf blades, but the scales subtending the achenes are distinctly *Fuirena*-like.

3. **Fuirena simplex** Vahl, Ecolog. Amer. 2:8. 1798. Fig. 140.
Fuirena squarrosa Michx. var. *aristulata* Torr. Ann. Lyc. N.Y. 3:291. 1798.
Fuirena simplex Vahl var. *aristulata* (Torr.) Kral, Sida 7:336. 1972.

Cespitose annual or more commonly a rhizomatous perennial; culms to 1 m tall, unbranched, puberulent at least in the upper half; leaves linear, 3–8 mm wide, with spreading pubescence; inflorescence with 1–4 headlike clusters of spikelets; spikelets lanceoloid to ovoid, up to 15 mm long; scales oblong to obovate, hispid, ciliate, 2.5–3.5 mm long, with a recurved awn 3–7 mm long; scaly bristles clawed at base and with a barbed awn at apex; stamens 1–3; anthers 0.5–1.2 mm long; style 3-cleft; achenes 0.8–1.1 mm long, trigonous, shiny, yellow to yellow-brown, stipitate. June–October.

Around ponds, along rivers, fens.

IA, MO (OBL); KY (NI); KS (not listed from KS by the U.S. Fish and Wildlife Service).

Umbrella sedge.

This species sometimes is a cespitose

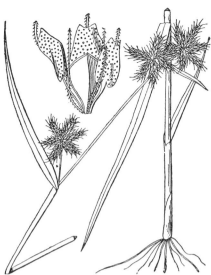

140. *Fuirena simplex.* Habit and spikelet.

annual, but usually it is a perennial with slender rhizomes. The annual plants may be designated var. *aristulata.* The plants are usually more than 60 cm tall, while the similar-appearing *F. annua* never is more than 60 cm tall.

10. **Hemicarpha** Nees & Arn.—Dwarf Bulsedge

Dwarf annuals; leaves 1 per culm, basal, capillary; inflorescence consisting of 1–several spikes subtended by 2–3 small involucral bracts, the longest appearing like a continuation of the culm; spikelets 1-flowered; each achene subtended by one large and one small scale; stamen 1; style 2-cleft; achenes compressed, papillate.

Approximately 6–8 species comprise this genus. The species of *Hemicarpha* are sometimes included in *Scirpus,* sometimes in *Lipocarpha.* I believe the differences are significant enough to justify genus status for *Hemicarpha.*

1. Scaly bristles subtending each achene shorter than the achene or absent; bristles often spreading and shorter than the body of the scale .... 3. *H. micrantha*
1. Scaly bristles subtending each achene appressed or, if spreading, as long as or longer than the body of the scale.
   2. Scales spreading to recurved, long-acuminate .... 1. *H. aristulata*
   2. Scales appressed, obtuse to subacute .... 2. *H. drummondii*

1. **Hemicarpha aristulata** (Coville) Smythe, Trans. Kansas Acad. Sci. 16:163. 1899. Fig. 141.
Hemicarpha micrantha (Vahl) Britt. var. *aristulata* Coville, Bull. Torrey Club 21:36. 1894.
Lipocarpha aristulata (Coville) G. C. Tucker, Journ. Arn. Arb. 68:410. 1987.

Dwarf cespitose annual to 10 cm tall; culms slender, glabrous; leaf 0.3–0.5 mm wide; inflorescence with 1–3 sessile spikelets; spikelets ovoid, 1–4 mm long; scales

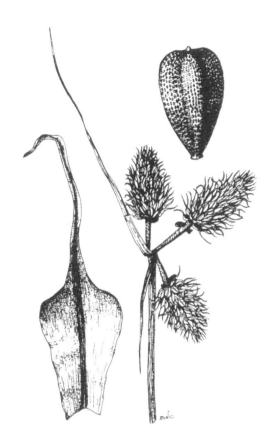

141. *Hemicarpha aristulata.* Habit, achene, and scale.

elliptic, with a long-tapering, spreading or recurved tip, the tip as long as or longer than the body of the scale, about as long as the achene; achenes lanceoloid, 0.5–0.8 mm long. July–October.

Moist depressions on glades, sometimes in shallow water.

MO. This species, which is called *Lipocarpha aristulata* by the U.S. Fish and Wildlife Service, is not listed for the north-central states, but it is OBL, FACW+, or FACW elsewhere.

Bristly dwarf bulsedge.

This dwarf plant differs from the other species in the genus by its recurved scales whose tip is as long as or longer than the body of the scale.

2. **Hemicarpha drummondii** Nees in Mart. Fl. Bras. 2 (1):61. 1842. Fig. 142.
Hemicarpha micrantha (Vahl) Pax var. *drummondii* (Nees) Friedl. Am. Journ. Bot. 28:860. 1941.
Scirpus micranthus Vahl. var. *drummondii* (Nees) Mohlenbr. Am. Midl. Nat. 70:22. 1963.
Lipocarpha drummondii (Nees) G. C. Tucker. Journ. Arn. Arb. 68:410. 1987.

Dwarf cespitose annual to 10 cm tall; culms slender, glabrous; leaf 0.3–0.5 mm wide; inflorescence with 1–3 sessile spikelets; spikelets more or less ovoid, 2–4 mm

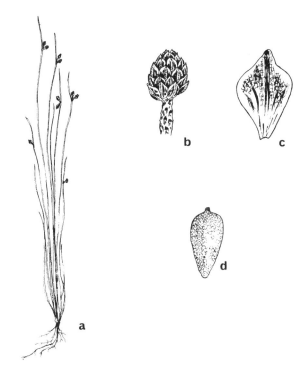

142. *Hemicarpha*
drummondii.

a. Habit.
b. Spikelet.

c. Scale.
d. Achene.

long; scales obovate to elliptic, appressed, bluntly tapered at tip, 0.9–1.3 mm long;
scalelike bristles about as long as the achene; achenes 0.3–0.5 mm long. July–October.

Around ponds, along rivers and streams, moist depressions, sandbars.

IL, IN, MO (FACW+); OH (OBL); KS, NE (FACW). This species is called *Lipocarpha*
drummondii by the U.S. Fish and Wildlife Service.

Drummond's dwarf bulsedge.

This species differs from the other two *Hemicarpha* species in the central Midwest
by its appressed scales.

3. **Hemicarpha micrantha** (Vahl) Pax, in Engl. & Prantl, Nt. Pflanz. 2 (2):105.
1887. Fig. 143.
Scirpus micranthus Vahl, Enum. Pl. 2:254. 1806.
Isolepis subsquarrosa Muhl. var. *minor* Schrad. in Roem. & Schult. Mant. 2:64. 1822.
Hemicarpha micrantha (Vahl) Pax var. *minor* (Schrad.) Friedl. Am. Journ. Bot.
28:860. 1941.
Lipocarpha micrantha (Vahl) G. C. Tucker, Journ. Arn. Arb. 68:410. 1987.

Dwarf cespitose annual to 10 cm tall; culms slender, glabrous; inflorescence
with 1–3 sessile spikelets; involucral leaf 1 (–3), very slender; spikelets more or less
ovoid, 2–4 mm long; scales narrowly obovate, spreading, mucronulate, glabrous,

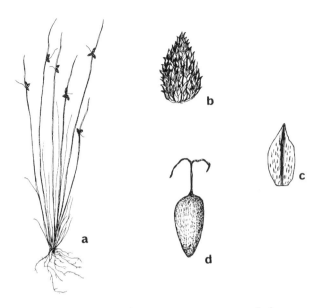

143. *Hemicarpha*
micrantha.

a. Habit.
b. Spikelet.

c. Scale.
d. Achene.

brown with a conspicuous green midvein, 0.3–0.8 mm long; achenes narrowly cylindric, remotely low-papillate, faintly reticulate, grayish brown, glabrous.

Around lakes and ponds, along rivers and streams, mud flats, sandbars, sloughs. IA, IL, IN, KS, KY, MO, NE, OH (OBL). The U.S. Fish and Wildlife Service calls this species *Lipocarpha micrantha.*

Common dwarf bulsedge.

Although not abundant, or perhaps overlooked because of its diminutive size, this is the most common species in the genus. The scales of the spikelets are tightly appressed.

11. **Lipocarpha** R. Br. in Tuckey—Lipocarpha

Tufted annuals (in the central Midwest) or perennials; leaves mostly near the base of the plant; inflorescence crowded, subtended by leafy bracts; spikelets many-flowered, terete; scales imbricate in several ranks, the lowermost sterile, minute, dilated, rudimentary; stamens 1 or 2; style 2- or 3-cleft; achenes lenticular or trigonous.

As considered in this work, *Lipocarpha* consists of 25 species. Some botanists merge *Hemicarpha* with *Lipocarpha*, a view not accepted here. The minute, dilated, rudimentary scales distinguish this genus from all other Cyperaceae.

Only the following species occurs in the central Midwest.

1. **Lipocarpha maculata** (Michx.) Torr. Ann. Lyc. N.Y. 3:288. 1836. Fig. 144. *Kyllinga maculata* Michx. Fl. Bor. Am. 1:29. 1803.

Dwarf annual with fibrous roots; culms to 25 cm tall; leaves short, linear, near the base of the plant; involucral leaves 2–5; spikelets ovoid-cylindrical, 3–7 mm

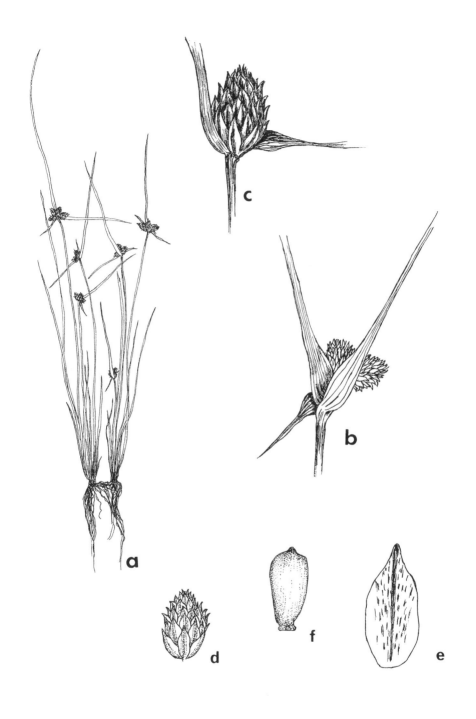

144. *Lipocarpha maculata.*
a. Habit.
b. Inflorescence.
c, d. Spikelets.
e. Scale.
f. Achene.

long; scales spatulate, imbricate, awnless, minute, dilated, rudimentary; stamen 1; style 2- to 3-cleft; achenes oblongoid, contracted to base. July–October.

Around a pond.

IL (OBL).

Lipocarpha.

This is a species of the Atlantic coastal plain with one disjunct location in central Illinois where it has not been seen in nearly half a century. Instead of being subtended by bristles, the achenes are subtended by minute, dilated, rudimentary scales.

12. Psilocarya Torr.—Bald Sedge

Annuals from fibrous roots; culms slender, leafy, more or less angular, glabrous; inflorescence a series of terminal and axillary umbels, the branches bracteate; spikelets many-flowered; stamens 1 or 2; style 2-cleft; bristles absent; achenes lenticular or biconvex, smooth or transversely wrinkled, with a tubercle persisting as a beak.

This genus consists of about 10 species, all confined to the new world. Because of the tubercle on the achene, some botanists place the species of this genus in *Rhynchospora*, but I am maintaining *Psilocarya* as a separate genus because of the absence of bristles and the many-flowered spikelets.

1. Tubercle of achene depressed, much shorter than the achene .... 1. *P. nitens*
1. Tubercle of achene subulate, nearly as long as the achene .... 2. *P. scirpoides*

1. **Psilocarya nitens** (Vahl) Wood, Am. Bot. Fl. 364. 1870. Fig. 145.
Scirpus nitens Vahl, Enum. 2:272–273. 1805.
Rhynchospora nitens (Vahl) Gray, Man.
Bot., ed. 5, 568. 1867.

Tufted annual; stems slender, to 60 cm tall; leaves 1–3 mm wide, glabrous; inflorescence cymose, consisting of several spikelets; spikelets ovoid to cylindric, up to 7 mm long; scales ovate, acute at the tip, about 3 mm long; bristles absent; style 2-cleft; stamens 1 or 2; achenes nearly globose, 0.7–1.0 mm in diameter, rugose, brown; tubercle very flat and reduced. June–September.

Bogs.

IN (OBL). This species is called *Rhynchospora nitens* by the U.S. Fish and Wildlife Service.

Psilocarya.

This is a coastal plain species with a disjunct station in Indiana.

145. *Psilocarya nitens.* Habit, spikelet, and achene.

Although some botanists place this species in *Rhynchospora*, I reject this view.

2. **Psilocarya scirpoides** Torr. Ann. Lyc. N.Y. 3:360. 1836. Fig. 146.
Rhynchospora scirpoides (Torr.) Griseb. Cat. Pl. Cub. 247. 1866.

Tufted annual; culms slender, to 60 cm tall; leaves 1–3 mm wide, glabrous; inflorescence cymose, consisting of several spikelets; spikelets ovoid to cylindric, up to 7 mm long; scales ovate, acute, about 3 mm long; bristles absent; style 2-cleft; stamens 1 or 2; achenes nearly globose, 0.7–1.0 mm in diameter; tubercle subulate, flat, nearly as long as the achene. June–September.

Wet soil; usually not in standing water

IN (OBL). The U.S. Fish and Wildlife Service calls this species *Rhynchospora scirpoides.*

Scirpus-like psilocarya.

This is an Atlantic coastal plain species with a few disjunct inland stations. It closely resembles *P. nitens* but has a subulate tubercle nearly as long as the achene. The tubercle in *P. nitens* is minute and depressed.

146. *Psilocarya scirpoides.* Habit, achene, and spikelet.

13. **Rhynchospora** Vahl—Beaked Sedge

Slender to robust perennials with bulbous bases; leaves capillary to linear; inflorescence composed of glomerules or spikelets; spikelets sessile or pedicellate; lowermost scales usually sterile, the uppermost subtending perfect or rarely imperfect flowers; bristles (0–) 3–14 (–20), shorter than to longer than the achenes; stamens usually 3; styles simple or 2-cleft, persistent as a beak upon the achene; ovary superior, 1-locular; achenes lenticular, rounded or flattened.

Rhynchospora is readily readily recognized by the conspicuous style persistent on the achene in the form of a beak.

As recognized in this work, there are about 225 species of *Rhynchospora* distributed worldwide.

1. Leaves 6–20 mm broad; achenes 4.7–5.5 mm long, 2.5–3.0 mm broad; beak of achene 12–20 mm long.
  2. At least one of the usually 6 bristles longer than the achene .... 8. *R. macrostachya*
  2. None of the usually 3–5 bristles longer than the achene .... 4. *R. corniculata*
1. Leaves 0.3–6.0 mm broad (rarely 7 mm in *R. glomerata*); achenes 1.2–2.6 mm long, 0.8–1.6 mm broad; beak of achene 0.3–0.6 mm long.
  3. Bristles antrorsely hairy, rarely smooth or entirely absent; tubercle 0.3–0.6 mm high.
    4. Achenes with transverse wavy ridges; tubercles deltoid .... 5. *R. globularis*
    4. Achenes honeycomb-reticulate; tubercles conical .... 7. *R. harveyi*
  3. Bristles retrorsely hairy, rarely smooth; tubercle 0.6–1.8 mm long.

5. Spikelets lanceoloid; bristles 3.6–4.3 mm long; leaves capillary, at most 0.4 mm broad .... 2. *R. capillacea*
5. Spikelets ovoid to lance-ovoid; bristles 2.0–3.5 mm long (occasionally to 4 mm long in *R. glomerata*); leaves 0.5–7.0 mm broad.
   6. Achenes with a conspicuous margin, smooth.
       7. Achenes 1.2–1.4 mm broad, with broad shoulders; bristles 3–4 mm long .... 6. *R. glomerata*
       7. Achenes 0.8–1.2 mm broad, pyriform; bristles 2.0–2.8 mm long .... 3. *R. capitellata*
   6. Achenes emarginate, or nearly so, slightly rugulose .... 1. *R. alba*

1. **Rhynchospora alba** (L.) Vahl, Enum. 2:236. 1806. Fig. 147.
Schoenus albus L. Sp. Pl. 1:44. 1753.

Cespitose perennial with very slender culms to 55 (–70) cm tall; leaves filiform to linear, 0.5–2.5 mm wide; glomerules whitish or light brown, the terminal one 5–15 mm broad, the lateral two somewhat smaller; spikelets sessile; achenes obovoid to pyriform, stipitate, slightly rugulose, lustrous, 1.5–2.0 mm long, the base about 1/2 as broad as the summit of the achene; beak elongate-triangular, 0.6–1.2 mm long, 1.2–1.8 mm broad; bristles usually 10–14, 2.4–3.0 mm long, retrorsely hairy. July–September.

Bogs, conifer swamps.
IL, IN, OH (OBL).
Pale beaked sedge.
This beaked sedge has retrorsely hairy bristles and smooth achenes. Its achenes lack a conspicuous margin.

2. **Rhynchospora capillacea** Torr. Fl. N. & Mid. States 55. 1823. Fig. 148.
Rhynchospora capillacea Torr. var. *leviseta* E. J. Hill, Am. Nat. 10:370. 1876.

Cespitose perennial with capillary culms to 40 cm tall; leaves filiform to linear, involute, to 0.4 mm wide; glomerules usually 2, the terminal with 2–10 spikelets, the lateral with 1–4 spikelets; spikelets sessile or nearly so, lanceoloid; achenes oblong-obovoid, stipitate, 1.7–2.4 mm long, 0.9–1.4 mm broad; beak 0.8–1.5 mm long, the base about two-thirds as wide as the summit of the achene; bristles 6, longer than the achene, retrorsely hairy or smooth. June–September.

Bogs, marshes, fens, stream banks, seeps.
IA, IL, IN, MO (OBL).
Slender beaked sedge.
The distinguishing features of *R. capillacea* are its retrorsely hairy bristles and its lanceoloid spikelets. Specimens with glabrous bristles are called var. *leviseta*.

3. **Rhynchospora capitellata** (Michx.) Vahl, Enum. 2:235. 1806. Fig. 149.
Schoenus capitellatus Michx. Fl. Bor. Am. 1:36. 1803.
Rhynchospora glomerata (L.) Vahl var. *minor* Britt. Trans. N.Y. Acad. Sci. 11:89. 1892.
Rhynchospora capitellata (Michx.) Vahl var. *minor* (Britt.) S. F. Blake, Rhodora 20:28. 1918.

Cespitose perennial with slender culms rarely to 1.2 m tall; leaves filiform to linear, 0.5–3.0 mm wide; terminal glomerule subsessile, the lateral pedunculate;

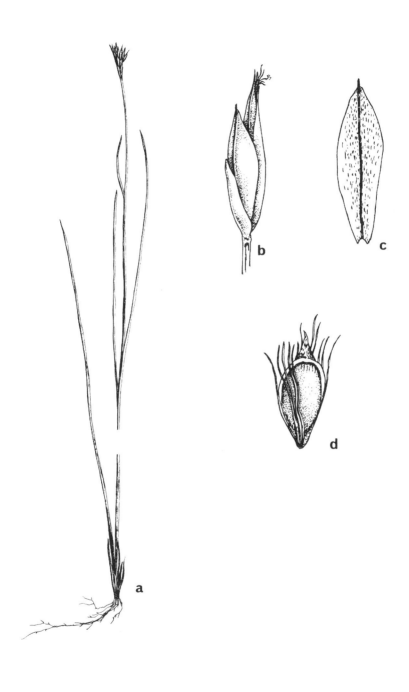

147. *Rhynchospora alba.*   a. Habit.   c. Scale.
                                            b. Spikelet.   d. Achene.

148. *Rhynchospora capillacea.*

a. Habit.
b. Spikelets.

c. Scale.
d. Achene.

spikelets 3–5 mm long; achenes pyriform, narrowly winged, 1.2–1.8 mm long, 0.8–1.2 mm broad; beak 1.0–1.5 mm long, the base abaout 1/2 as broad as the summit of the achene; bristles 5–6, retrorsely hairy or rarely glabrous, 2.0–2.8 mm long. May–October.

Fens, sinkhole ponds, along rivers and spring branches, in moist depressions. IL, MO, OH (OBL).

False bog sedge.

This species is very similar in appearance to *R. glomerata,* differing by its narrow achenes and shorter bristles.

149. *Rhynchospora capitellata.*

a. Habit.
b. Spikelet.

c. Scale.
d. Achene.

4. **Rhynchospora corniculata** (Lam.) Gray, Ann. Lyc. N.Y. 3:205. 1835. Fig. 150.
Schoenus corniculatus Lam. Tabl. Encycl. 1:137. 1791.
Rhynchospora corniculata (Lam.) Gray var. *interior* Fern. Rhodora 20:140. 1918.

Robust perennial with swollen bases; culms stout, erect, to nearly 2 m tall; leaves several, to 20 mm broad, with smooth or scaberulous margins; inflorescence much branched, often nearly 1 m long; glomerules with 4–10 flowers; achenes 4.7–5.5 mm long, 2.6–3.0 mm broad; beak 12–18 mm long, the base two-thirds to nearly equaling the width of the summit of the achene; bristles 3–5, 2–4 mm long, shorter than the achene. June–September.

Ditches, swampy woods, along spring branches.

IL, IN, KY, MO (OBL).

Horned sedge.

This large species, whose culms may reach 2 meters tall, differs from *R. macrostachya*, the other tall *Rhynchospora* in the central Midwest, by its bristles that are only 2–4 mm long. Both species have extremely long tubercles.

5. **Rhynchospora globularis** (Chapm.) Small var. **recognita** Gale, Rhodora 46:245. 1943. Fig. 151.
Rhynchospora cymosa var. *globularis* Chapm. Fl. So. U.S. 525. 1860.

Tufted perennial with slender, glabrous culms to 50 cm tall; leaves 2–5 mm broad, the sheaths glabrous; inflorescence cymose, the terminal glomerules 5–10 mm broad; spikelets crowded, 2.5–4.0 mm long; scales suborbicular, obtuse, occasionally mucronulate; achenes obovoid to suborbicular, honeycomb-reticulate, frequently rugulose, 1.3–1.6 mm long, slightly narrower; beak merely deltoid, 0.3–0.6 mm long; bristles tiny, much shorter than the achenes, antrorsely hairy. July–September.

Along streams, in sinkhole ponds.

IL, IN, KS, MO, OH (FACW).

Round-headed beaked sedge.

This species differs from others in the genus except *R. harveyi* by its antrorsely hairy bristles. It differs from *R. harveyi* by its slightly larger achenes with a honeycomb-reticulate surface.

6. **Rhynchospora glomerata** (L.) Vahl, Enum. 2:234. 1806. Fig. 152.
Schoenus glomeratus L. Sp. P. 1:441. 1753.

Cespitose perennial with rather slender culms to 1.5 m tall; leaves linear to linear-lanceolate, 3–6 (–7) mm broad; spikelets subsessile, 4.5–6.5 mm long; achenes suborbicular, with broad shoulders and a prominent margin, lustrous, 1.5–1.8 mm long, 1.2–1.4 mm broad; beak 1.2–1.8 mm long, the base about 1/2 as broad as the summit of the achene; bristles usually 6, retrorsely hairy, 3–4 mm long. June–September.

Moist soil, rarely in shallow water.

IL (OBL).

Clustered beaked sedge.

150. *Rhynchospora corniculata.*

a. Habit.
b. Spikelet.

c. Scale.
d. Achene.

151. *Rhynchospora*
globularis.

a. Habit.
b. Spikelets.

c. Scale.
d. Achene.

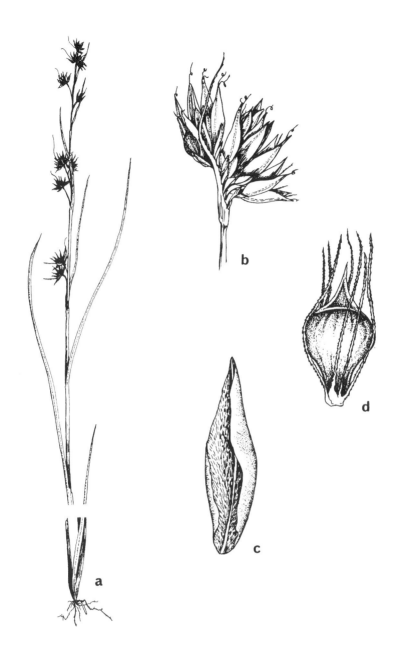

152. *Rhynchospora glomerata.*

a. Habit.
b. Spikelets.

c. Scale.
d. Achene.

153. *Rhynchospora harveyi.* Habit and achene.

This species differs from the very similar appearing *R. capitellata* by its wider achenes and longer bristles. It is far less common in the central Midwest than *R. capitellata.*

7. **Rhynchospora harveyi** W. Boott, Bot. Gaz. 9:85–86. 1884. Fig. 153.

Tufted perennial; culms triangular, to 60 (–65) cm tall; leaves 2.5–3.5 mm broad, flat; inflorescence a cluster of small glomerules, each glomerule consisting of a few spikelets; spikelets ovoid to broadly ellipsoid, obtuse, usually 2-flowered; scales ovate to broadly elliptic, awn-tipped, 1.8–2.0 mm long; bristles 5–8, shorter than to about as long as the achene, antrorsely hairy; achenes obovoid to subglobose, 1.5–1.8 mm long, honeycomb-reticulate; beak depressed-triangular, 0.3–0.6 mm long. June–July.

Wet depressions, sometimes in shallow water.

KS, MO (FAC).

Harvey's beaked sedge.

Although this species is given a FAC rating by the U.S. Fish and Wildlife Service, it sometimes occurs in standing water, particularly in Kansas.

This species has antrorsely hairy bristles as does *R. globularis*, but the achenes of *R. harveyi* have a honeycomb-reticulate surface. The surface of the achenes of *R. globularis* has transverse ridges.

8. **Rhynchospora macrostachya** Torr. ex A. Gray, Ann. Lyc. N.Y. 3:206. 1835. Fig. 154.

Robust perennial with bulbous-thickened bases; culms to 1 m tall, glabrous; leaves 6–12 mm broad, glabrous, the sheaths glabrous; inflorescence 0.6–1.0 m long, stiffly ascending, the terminal glomerules sessile or on stalks up to 4 mm long; spikelets 7–45 in a glomerule, reddish brown, 1.5–2.8 mm long; scales elliptic-lanceolate, subacute; achenes obovoid, flattened on the faces, short-stipitate, 4.7–5.4 mm long, 2.5–3.0 mm broad; beak subulate, serrulate, 1.5–2.0 cm long, 1–2 mm broad at base; bristles (4–) 6, antrorsely hairy, at least some of them 1 cm long or longer. July–October.

Around sinkholes, swampy woods, ditches.

IN, KS, MO (OBL).

Smaller horned sedge.

The long bristles distinguish this species from *R. corniculata*, the other species of *Rhynchospora* in the central Midwest with broad leaves.

14. **Schoenoplectus** (Reichenb.) Palla—Bulsedge

Tufted annuals or rhizomatous perennials; culms unbranched, terete or 3-angled; leaves basal or none, the 1–3 lowermost reduced to bladeless sheaths, glabrous; inflorescence lateral, composed of 1–many spikelets, some sessile and glomerulate, others in umbels; bracts erect and appearing like a continuation of the culm; spikelets 10- to many-flowered; scales usually short-awned; bristles 3–6, or absent; stamens 2 or 3; styles 2- or 3-cleft; achenes lenticular or trigonous.

The species of this genus traditionally have been included in *Scirpus*, but the lateral inflorescence is distinctive. There are 40 species of *Schoenoplectus* worldwide.

1. Culms 3-angled.
   2. Some or all of the spikelets pedicellate .... 5. *S. etuberculatus*
   2. All spikelets sessile.
      3. Annuals without rhizomes; lowest leaves all reduced to basal sheaths ....
      .... 8. *S. mucronatus*
      3. Perennials, often with rhizomes; usually at least 1 or more leaves at base of culm with a blade.
         4. Scales of spikelets not notched at tip, several-nerved .... 16. *S. torreyi*
         4. Scales of spikelets notched at tip, 1-nerved.
            5. Sides of triangular culm concave .... 2. *S. americanus*
            5. Sides of triangular culm flat.
               6. Perianth bristles shorter than to as long as the achene; plants up to 1.5 m tall
               .... 10. *S. pungens*
               6. Perianth bristles longer than the achene; plants often 2 m tall or taller ....
               .... 4. *S. deltarum*

154. *Rhynchospora*
macrostachya.

a. Habit.
b. Spikelet.

c. Scale.
d. Achene.

1. Culms terete or sometimes more or less flattened, not 3-angled.
  7. Some or all of the spikelets pedicellate.
    8. Achenes trigonous; all spikelets pedicellate .... 7. *S. heterochaetus*
    8. Achenes biconvex; at least some of the spikelets sessile.
      9. Basal sheaths markedly fibrillose on the margins; achenes subtended by 2–4
        bristles .... 3. *S. californicus*
      9. Basal sheaths smooth along the margins or with only a few fibers; achenes
        subtended by (4–) 6 bristles.
        10. Scales viscid, reddish-dotted, mucronulate .... 1. *S. acutus*
        10. Scales not viscid, not reddish-dotted, acute .... 15. *S. tabernaemontani*
  7. All spikelets sessile or nearly so.
    11. Perennial with rhizomes; achenes 2.2–3.8 mm long.
      12. Achenes trigonous; scales 4–6 mm long .... 14. *S. subterminalis*
      12. Achenes lenticular; scales 3–4 mm long.
        13. Scales red-punctate, viscid; bristles 4–6 .... 1. *S. acutus*
        13. Scales brown, not viscid; bristles 1–3 .... 9. *S. nevadensis*
    11. Tufted annuals (if rhizomes present, they are hidden); achenes 1.3–1.7 mm long.
      14. Achenes smooth .... 13. *S. smithii*
      14. Achenes transversely corrugated or pitted.
        15. Achenes transversely corrugated; scales awned or cuspidate.
          16. Scales awned; styles 2-cleft; achenes lenticular .... 6. *S. hallii*
          16. Scales cuspidate; styles 3-cleft; achenes trigonous .... 12. *S. saximontanus*
        15. Achenes pitted; scales obtuse .... 11. *S. purshianus*

1. **Schoenoplectus acutus** (Muhl.) A. Love & D. Love, Bull. Torrey Club 81:33.
1954. Fig. 155.
Scirpus glaucus Smith, Engel. Bot. 33, t. 2321. 1811, non Lam. (1791).
Scirpus acutus Muhl. ex Bigel. Fl. Bost. 15. 1814.
Scirpus lacustris L. ssp. *glaucus* (Smith) Hartm. Svensk. Norsk. Exc. FL. 10. 1846.
Scirpus lacustris L. var. *occidentalis* S. Wats. Bot. Calif. 2:218. 1880.
Scirpus occidentalis (S. Wats.) Chase, Rhodora 6:68. 1904.
Scirpus occidentalis Chase var. *congestus* Farw. Rep. Mich. Acad. 19:247. 1917.
Scirpus acutus Muhl. f. *congestus* (Farw.) Fern. Rhodora 23:131. 1921.
Scirpus malheurensis Henderson, Rhodora 32:20. 1930.
Schoenoplectus acutus (Muhl.) A. Love & D. Love f. *congestus* (Farw.) Mohlenbr. Ill. Fl.
Il. Sedges, ed. 2, 201. 2001.

Perennial with firm, usually dark brown rhizomes; culms terete, rather stiff,
erect, to 1.3 m tall; basal leaves absent, the sheaths fibrillose at maturity; inflorescence a stiff panicle with divergent rays up to 5 cm long, or occasionally all spikelets
sessile and forming a glomerule; spikelets narrowly ovoid to linear-cylindric, acute,
up to 1.7 cm long, reddish brown; scales narrowly ovate, acute, mucronulate, 4 mm
long, red-punctate, more or less villous; achenes black, lustrous, 2.0–2.5 mm long.
May–September.

Banks of rivers, around ponds and lakes, sloughs, ditches, marshes.

IA, IL, IN, KS, KY, MO, NE, OH (OBL). The U.S. Fish and Wildlife Service calls this
plant *Scirpus acutus*.

Hardstem bulsedge.

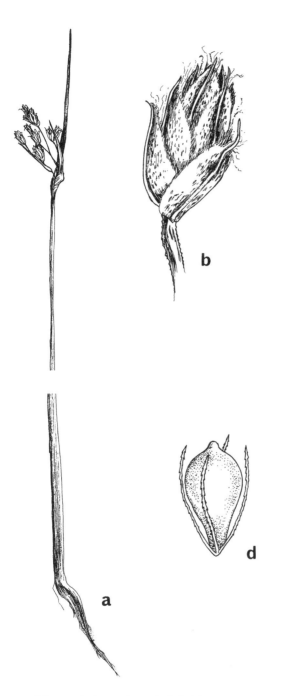

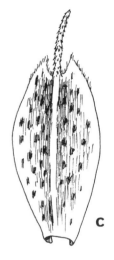

155. *Schoenoplectus acutus.*
a. Habit.

b. Spikelets.
c. Scale.

d. Achene.

a–d are f. *acutus;* e–f are f. *congestus.*

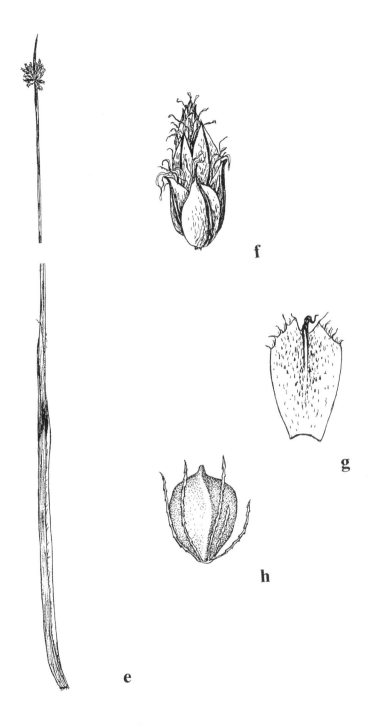

e. Habit.
f. Spikelet.

g. Scale.
h. Achene.

Rare specimens in which all the spikelets are sessile may be called f. *congestus*.

Scirpus acutus is distinguished by a combination of its terete culms, lenticular achenes, and reddish-punctate, viscid scales.

2. **Schoenoplectus americanus** (Pers.) Volkert ex Schinz & R. Keller, Fl. Schweiz. ed. 2, 1:75. 1905. Fig. 156.
Scirpus americanus Pers. Syn. 1:68. 1805.
Scirpus olneyi Gray, Bost. Journ. Nat. Hist. 5:238. 1845.

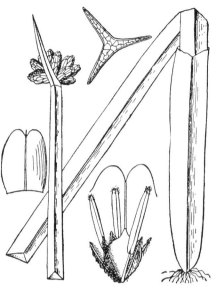

Perennial from thick, long-creeping rhizomes; culms erect, up to 1.5 m tall, stiff, strongly 3-angled with concave sides; leaves 2–4 near the base of the plant, the 1–3 lowermost leaves reduced to sheaths; inflorescence consisting of 2–15 sessile spikelets in a headlike cluster, subtended by a single, erect bract that appears as the continuation of the culm; spikelets ovoid, obtuse to subacute at the tip, several-flowered; scales broadly ovate, notched at the tip, short-awned, usually orange-brown tinged with purple; bristles 3–6, shorter than to a little longer than the achenes; styles 2-cleft; achenes lenticular, obovoid,

156. *Schoenoplectus americanus.* Habit, cross-section of stem, scale, and flower.

1.8–2.5 mm long, yellow to dark brown, shiny. June–September.

Alkaline seeps, often in standing water.

KS, MO, OH (OBL). This species is called *Scirpus americanus* by the U.S. Fish and Wildlife Service.

Saltmarsh bulsedge.

For a long time, this species was known as *Scirpus olneyi*, and the common plant of the Midwest previously known as *Scirpus americanus* is now *Schoenoplectus pungens*. *Schoenoplectus americanus* is rare in the central Midwest.

3. **Schoenoplectus californicus** (C. A. Meyer) Sojak, Cas. Nar. Mus., Odd. Prir. 140 (3–4):127. 1972. Fig. 157.
Elytrospermum californicum C. A. Meyer, Mem. Acad. Imp. Sci. St. Petersbourg, 6th series, 1:201, pl. 1, f. 2. 1830.
Scirpus californicus (C. A. Meyer) Steud. Nomen. Bot., ed. 2, 2:538. 1841.

Robust perennial from thickened bases; culms to 3 m tall, terete or obtusely angled, soft, glabrous; basal sheaths markedly fibrillose along the margins; bract 1, terete, appearing as a continuation of the culm, up to 7 cm long; panicle much branched, pendulous, bearing up to 150 spikelets; spikelets lance-ovoid, 6–11 mm long, up to 50-flowered; scales ovate to obovate, dark brown, mucronate, about 3 mm

157. *Schoenoplectus californicus.*   a. Habit.   b. Achene with bristles.

long, longer than the achenes; bristles 2–4, reddish brown with retrorse hairs; styles 2-cleft; achenes obovoid, plano-convex, brown, 2.0–2.2 mm long.

Around a pond, occasionally in shallow water.

IL. This species is not indicated for the Midwest by the U.S. Fish and Wildlife Service, but it is OBL elsewhere. The U.S. Fish and Wildlife Service calls this species *Scirpus californicus.*

Giant bulsedge.

This species, introduced from the western United States, occurs around a pond in Illinois. It resembles *S. tabernaemontani,* differing by its extremely fibrillose basal sheaths and its smaller achenes.

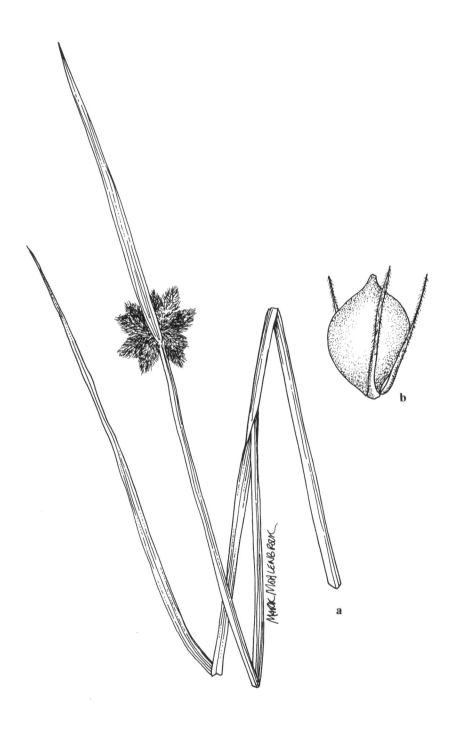

158. *Schoenoplectus deltarum.*

a. Habit.

b. Achene with bristles.

4. Schoenoplectus deltarum (Schuyler) Sojak, Cas. Nar. Mus., Odd. Prir. 141:62. 1972. Fig. 158.
Scirpus deltarum Schuyler, Not. Nat. Acad. Sci. Phila. 427:1. 1970.

Stout perennial from thickened rhizomes; culms to 2.5 m tall, strongly 3-angled, scabrous along the angles; lowest sheaths often bladeless, U-shaped at the summit, or sometimes 1–2 sheaths with blades, the blades very thick, up to 1 cm broad; involucral bract 1, appearing as a continuation of the culm, to 18 cm long; inflorescence of up to 12 crowded, sessile spikelets; spikelets ovoid to narrowly ellipsoid, mostly acute at the tip, up to 15 mm long; scales ovate, notched at the tip and mucronate or even short-awned, 3.5–6.0 mm long, longer than the achenes; bristles 3–6, slender, curved, retrorsely hairy; achenes 2.0–2.5 mm long, obovoid, biconvex, smooth, yellow to dark brown, shiny. July–October.

Ditches, marshy areas, occasionally in shallow water.

IL, MO (NI). In other regions, the U.S. Fish and Wildlife Service designates this species either FACW+ or OBL. This species is called *Scirpus deltarum* by the U.S. Fish and Wildlife Service.

Delta bulsedge.

This species most nearly resembles *S. pungens* but is more robust, has more spikelets per inflorescence, and has much longer and more slender bristles.

5. Schoenoplectus etuberculatus (Steud.) Sojak, Cas. Nar. Mus., Odd. Prir. 140 (3–4):127. 1972. Fig. 159.
Rhynchospora etuberculata Steud. Syn. Pl. Glum. 2:143. 1855.
Scirpus etuberculatus (Steud.) Fern. Rhodora 8:162. 1906.

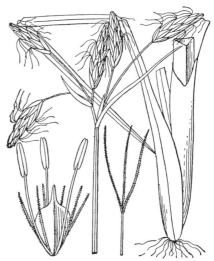

Perennial from slender, long-creeping rhizomes; culms erect, to 2 m tall, soft, 3-angled, glabrous, blue-green to gray-green; leaves usually 3–10, basal, to 1.5 m long; sheaths usually fibrillose at the tip; inflorescence of up to 15 spikelets, most of them on peduncles of irregular lengths forming an umbel; main involucral bract 1, erect, to 20 cm long, with several small pbracts subtending the branches of the inflorescence; spikelets 10–20 mm long, ovoid, acute; scales 5–6 mm long, ovate, not notched at tip,

159. *Schoenoplectus etuberculatus.*
Habit, spikelet, flower, and stigmas.

yellow-brown except for the green midvein; bristles usually 6, shorter than to slightly longer than the achenes, antrorsely hairy; styles 3-cleft; achenes 3.0–4.5 mm long, obovoid but tapering to a narrow beak, trigonous, smooth, yellow to black, shiny. June–October.

160. *Schoenoplectus hallii.*
a. Habit.
b. Spikelets.
c. Scale.
d, e. Achenes.

On floating mats in a sinkhole pond.

MO (OBL). The U.S. Fish and Wildlife Service lists this species as *Scirpus etuberculatus.* Canby's bulsedge.

This Atlantic coastal plain species has a single disjunct station in the Missouri Ozarks. It differs by having all the spikelets pedicellate. The culms have a distinctly blue-green or gray-green color.

6. **Schoenoplectus hallii** (Gray) S. G. Sm. Novem 5 (1):101. 1995. Fig. 160.
Scirpus hallii Gray, Man. Bot., ed. 3, addend. xvii. 1863.
Scirpus supinus L. var. *hallii* Gray, Man. Bot., ed. 5, 563. 1867.
Scirpus uninodis var. *hallii* (Gray) Beetle, Am. Journ. Bot. 29:656. 1942.

Slender, tufted annual; culms terete, to 35 cm long, with several short culms near base of plant; basal sheaths with or without a blade; involucral leaf 1, acute, nearly 1/2 as long as the culm; spikelets 1–7, ovoid-cylindric, acute, 5–15 mm long, 2.5–3.5 mm broad, greenish brown; scales ovate, acuminate and with a cusp, 2.5–3.0 mm long, green and brown; bristles absent; achenes obovoid, plano-convex, transversely corrugated, 1.5–2.0 mm long, black at maturity, usually in spikelets but sometimes at the base of the culm. July–October.

Shore of ponds, in sinkhole ponds, depressions in sandy prairies.

IA, IL, IN, KS, MO, NE (OBL). This species is called *Scirpus hallii* by the U.S. Fish and Wildlife Service.

Hall's bulsedge.

This species has distinctive black, transversely corrugated achenes. It also possesses some achenes which are borne at the base of the culm rather than in spikelets.

7. **Schoenoplectus heterochaetus** (Chase) Sojak, Cas. Nar. Mus., Odd. Prir. 140 (3–4):127. 1972. Fig. 161.
Scirpus heterochaetus Chase, Rhodora 6:70. 1904.

Perennial from firm rhizomes; culms slender, erect, to 1 m tall, terete, glabrous; inflorescence paniculate, with very slender and nearly glabrous branchlets; involucral bract 1, appearing as a continuation of the stem, up to 3 cm long; spikelets all pedicellate, 12–22 mm long, lanceoloid to ellipsoid, acute to acuminate, pale brown or whitish; scales ovate, acute, cleft at the apex, occasionally red-dotted, glabrous, about 4 mm long; bristles usually 2, delicate, not surpassing the achenes; achenes trigonous, 2.2–2.5 mm long, brownish, smooth. May–September.

Ditches, sloughs, swamps, marshes, around ponds and lakes, along rivers, in borrow pits, often in shallow water.

IA, IL, IN, KS, KY, MO, NE (OBL). The U.S. Fish and Wildlife Service calls this species *Scirpus heterochaetus.*

Slender bulsedge.

This species resembles *S. acutus* and *S. tabernaemontani.* The following chart compares and contrasts these three species.

S. heterochaetus	*S. acutus*	*S. tabernaemontani*
panicle branches flexuous	panicle branches stiff	panicle branches stiff
spikelets 12–22 mm long	spikelets 10–17 mm long	spikelets 5–9 mm long
scales glabrous	scales viscid	scales pubescent at tip
scales sometimes red-dotted	scales red-dotted	scales not dotted

8. **Schoenoplectus mucronatus** (L.) Palla ex Kerner, Sched. Fl. Exs. Austro-Hung. 5:91. 1888. Fig. 162.
Scirpus mucronatus L. Sp. Pl. 1:50. 1753.

Tufted annual from fibrous roots; culms erect, to 75 cm long, stiff, strongly 3-angled, usually more or less scabrous on the angles; lowest sheaths always without blades, usually V-shaped at the summit; involucral bract 1, appearing as a continuation of the culm, up to 12 cm long, usually bent to one side; inflorescence com-

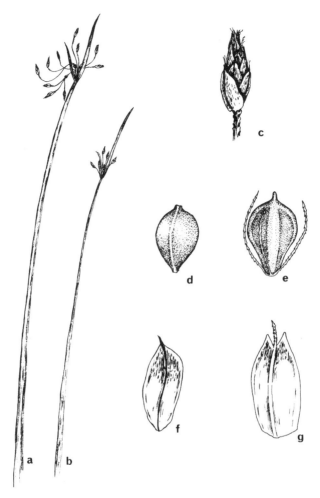

161. *Schoenoplectus
heterochaetus.*

a, b. Habit.
c. Spikelet.

d, e. Achenes.
f, g. Scales.

posed of up to 10 sessile spikelets; spikelets ovoid to ellipsoid, more or less obtuse at the tip, up to 12 mm long; scales obovate to nearly rotund, obtuse but mucronate, chartaceous, 2.5–3.5 mm long; bristles 6, as long as or longer than the achenes; styles 3-cleft; achenes more or less trigonous, 1.7–2.0 mm long, transversely wrinkled, green to dark brown, shiny. June–September.

Around lakes and ponds, along rivers, usually in shallow water.

IA, IL, MO (NI). In other regions, the United States Fish and Wildlife Service designates this species as OBL and lists it as *Scirpus mucronatus.*

Mucronate bulsedge.

This native of Europe and Asia grows as an adventive in the central Midwest. It is the only annual *Schoenoplectus* with triangular culms.

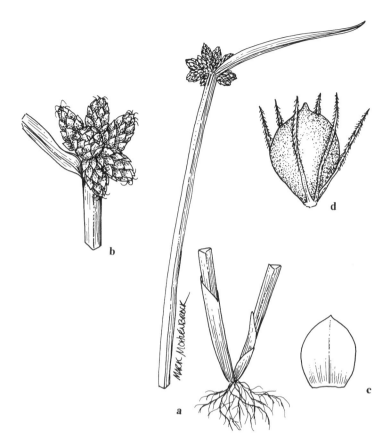

162. *Schoenoplectus*
mucronatus.

a. Habit.
b. Spikelets.

c. Achene with bristles.
d. Scale.

9. **Schoenoplectus nevadensis** (S. Wats.) Sojak, Cas. Nar. Mus., Odd. Prir. 140 (3–4):127. 1972. Fig. 163.
Scirpus nevadensis S. Wat. U.S. Geol. Expl. 360. 1871.

Perennial from tough rhizomes; culms erect, slender, terete, to 40 cm tall, glabrous; leaves few, basal, up to 2 mm broad; inflorescence consisting of compact clusters of spikelets; main involucral bract 1, appearing as a continuation of the culm, with smaller bracts subtending the individual spikelets; spikelets sessile, ovoid to oblongoid, 12–18 mm long, sessile; scales ovate, acute, brown; bristles 1–3; styles 2-cleft; achenes lenticular, minutely reticulate. July–October.

Alkaline depressions, sometimes in shallow water.

NE (OBL). The U.S. Fish and Wildlife Service calls this species *Scirpus nevadensis*. Nevada bulsedge.

This is a western species that is distinguished by its terete stems and few (1–3) bristles.

JRJ

163. *Schoenoplectus nevadensis.* Two habit sketches, scale, achene, and spikelets with bracts.

10. **Schoenoplectus pungens** (Vahl) Palla, Verh. Zool.-Bot. Ges. Wien 38:49. 1888. Fig. 164.
Scirpus americanus Pers. Syn. 1:68. 1805, misapplied.
Scirpus pungens Vahl, Enum. Pl. 2:255. 1806.

Perennial from a stout, deep brown rhizome; culms erect, numerous, 3-angled, to 1.5 m tall, glabrous; lowest sheaths bearing blades, the blades linear, up to 9 mm broad; inflorescence a small cluster of sessile spikelets; involucral bract 1, erect, appearing as a continuation of the culm, slender, acute, up to 15 cm long; spikelets 1–8, ovoid or cylindric, subacute at the apex, 5–18 (–24) mm long; scales ovate, 2-cleft at the apex, usually ciliate, otherwise glabrous; bristles several, stout, short, usually persistent; achenes obovoid, plano-convex, smooth, pale brown to olive, 2.5–3.0 mm long. May–September.

Sloughs, ditches, marshes, fens, around ponds and lakes, along rivers and streams, often in shallow water.

IA, IL, IN, MO (NI), KS, NE (OBL), KY, OH (FACW+). This species is called *Scirpus pungens* by the U.S. Fish and Wildlife Service.

Three-square.

In the past, this species has been called *Scirpus (=Schoenoplectus) americanus*, but that binomial applies to a different species. *Schoenoplectus pungens* is distinguished by its triangular stems, notched scales of the spikelets, and usually only 1–4 (–8) spikelets in a cluster.

11. **Schoenoplectus purshianus** (Fern.) M. T. Strong, Novon 3 (2):202. 1993. Fig. 165.
Scirpus debilis var. *williamsii* Fern. Rhodora 3:252. 1901.
Scirpus smithii Gray var. *williamsii* (Fern.) Beetle, Am. Journ. Bot. 21:655. 1942.
Scirpus purshianus Fern. Rhodora 44:479. 1942.
Scirpus purshianus Fern. f. *williamsii* (Fern.) Fern. Rhodora 44:479. 1942.
Scirpus juncoides Roxb. var. *williamsii* (Fern.) Koyama, Can. Journ. Bot. 40:914. 1962.

Rather coarse, tufted annual; culms stout, erect, 3-angled, soft, to 50 cm tall, glabrous; basal sheaths usually bladeless; inflorescence composed of 1–6 sessile spikelets in a headlike cluster; involucral bract 1, appearing as a continuation of the culm, erect, rounded at the tip, one-fourth to one-sixth as long as the culm; spikelets ovoid or cylindric, obtuse at the tip, 5–10 mm long, 3–5 mm broad, with rather pale margins; bristles (0–) 3–several, stout, persistent, with at least 2–3 of them surpassing the achene; styles 2-cleft; achenes obovoid, flattened, lenticular, pitted.

Around lakes, often in shallow water.

IL, IN, MO, OH (OBL). The U.S. Fish and Wildlife Service calls this species *Scirpus purshianus*.

Pursh's bulsedge.

This species has pitted, lenticular achenes, stout bristles, and obtuse scales.

12. **Schoenoplectus saximontanus** (Fern.) J. Raynal, Adansonia n.s. 16 (1):141. 1976. Fig. 166.
Scirpus saximontanus Fern. Rhodora 3:251. 1901.
Scirpus supinus L. var. *saximontanus* (Fern.) Koyama, Can. Journ. Bot. 40:921. 1962.

164. *Schoenoplectus*
pungens.

a. Habit.
b. Spikelets.

c, d. Scales.
e. Achene.

a

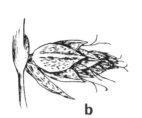

b

c

d

165. *Schoenoplectus*
purshianus.

a. Habit.
b. Spikelet.

c. Scale.
d. Achene.

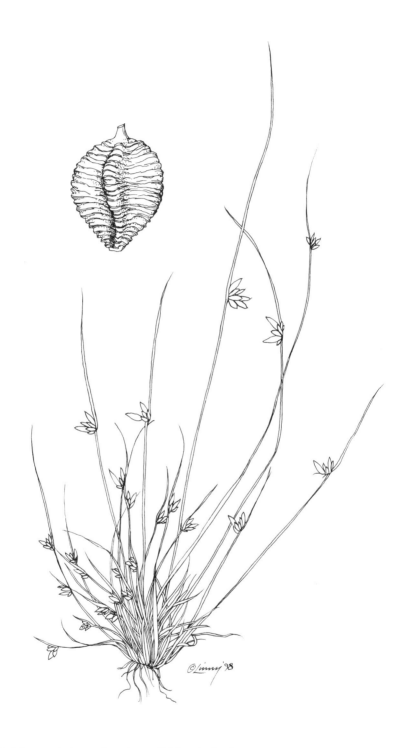

166. *Schoenoplectus saximontanus.* Habit and achene.

Tufted annual; culms terete to slightly flattened, erect, to 40 cm tall, glabrous; leaves 2–3 near base of plant, all others reduced to bladeless sheaths; inflorescence a headlike cluster of 1–10 spikelets; involucral bracts 2, one long appearing as a continuation of the culm, and one short; spikelets ovoid to lanceoloid, obtuse to acute at the tip, 5–15 mm long; scales ovate, acute, cuspidate, 2.5–4.0 mm long, green and usually orange-brown; bristles usually absent; styles 3-cleft; achenes obovoid to broadly ellipsoid, trigonous, 1.3–1.7 mm long, dark brown to black, with wavy, transverse ridges. July–September.

Depressions in marshes, ditches, around ponds, sometimes in shallow water.

KS, NE, OH (OBL), MO (NI). This species is listed as *Scirpus saximontanus* by the U.S. Fish and Wildlife Service.

Rocky Mountain bulsedge.

Sometimes a few achenes are formed in the basal sheaths of the plant, similar to the condition found in *S. hallii*. *Schoenoplectus saximontanus* differs from all annual species of *Schoenoplectus* that have terete culms by its 3-cleft styles and its trigonous achenes.

13. **Schoenoplectus smithii** (Gray) Sojak, Cas. Nar. Mus., Odd. Prir. 141:62. 1972. Fig. 167.
Scirpus smithii Gray, Man. Bot., ed. 5, 563. 1861.

Tufted annuals; culms slenter, terete, subangular, to 35 cm tall, glabrous; lowest sheaths blade-bearing or bladeless; inflorescence a headlike cluster of 1–10 sessile spikelets; involucral bract 1, appearing as a continuation of the culm, erect, acute, nearly 1/2 as long as the culms; spikelets ovoid, acute, 5–10 mm long, 3–5 mm broad, greenish; scales oblong-ovate to ovate, obtuse, frequently mucronulate, greenish brown; bristles absent or few and slender; achenes obovoid, flat on one surface, slightly convex on the other, smooth, 1.8–2.0 mm long, black, lustrous. July–September.

Around lakes and ponds, occasionally in shallow water.

IA, IL, IN, NE (OBL). The U.S. Fish and Wildlife Service calls this species *Scirpus smithii*.

Smith's bulsedge.

The smooth achenes distinguish this species from the similar appearing *S. hallii* and *S. purshianus*.

14. **Schoenoplectus subterminalis** (Torr.) Sojak, Cas. Nar. Mus., Odd. Prir. 140 (3–4):127. 1972. Fig. 168.
Scirpus subterminalis Torr. Fl. U.S. 1:47. 1824.

Perennial from soft, slender rhizomes; culms erect, capillary, to 1 m tall, glabrous; leaves numerous, capillary; inflorescence consisting of a single, sessile, lateral spikelet; involucral bract 1, appearing as a continuation of the culm, erect, acute at the tip, filiform, to 6 mm long; spikelet lanceoloid to ovoid, to 12 mm long, to 7 mm broad, pale greenish brown; scales ovate-lanceolate, 4–6 mm long, pale brown with a green midrib; bristles 2–3 mm long, retrorsely hairy; achenes obovoid, short-beaked, 2.7–3.2 mm long, olive-brown, lustrous.

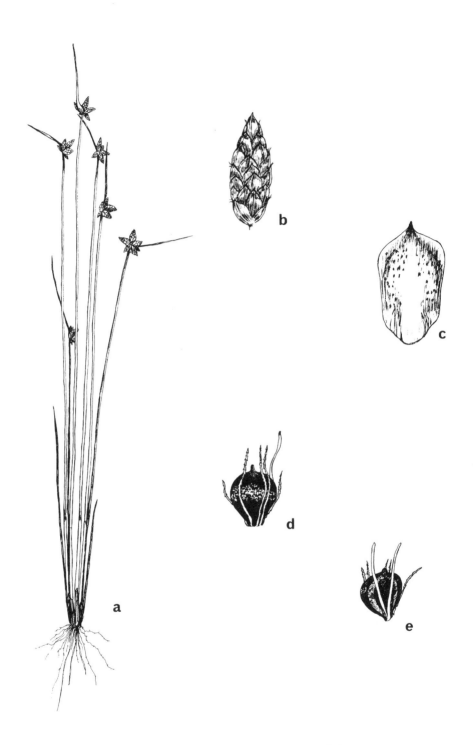

167. *Schoenoplectus smithii.*   a. Habit.
                                   b. Spikelet.

c. Scale.
d, e. Achene.

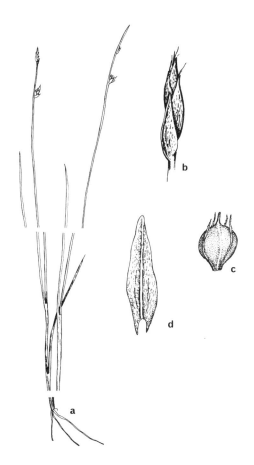

168. *Schoenoplectus*
subterminalis.

a. Habit.
b. Spikelet.

c. Achene.
d. Scale.

In sinkhole ponds, in deep water of streams.

IL, IN, MO (OBL). This species is called *Scirpus subterminalis* by the U.S. Fish and Wildlife Service.

Swaying bulsedge.

Of the species of *Schoenoplectus* with terete culms and sessile spikelets, this is the only one that has only a single spikelet per culm and capillary culms. Its achenes are larger than those of *S. hallii, S. purshianus, S. saximontanus,* and *S. smithii.*

15. **Schoenoplectus tabernaemontani** (C. C. Gmel.) Palla, Sitzb. Zool.-Bot. Ges. Wien 38:49. 1888. Fig. 169.
Scirpus tabernaemontani C. C. Gmel. F. Bad. 1:101. 1805.
Scirpus validus Vahl, Enum. Pl. 2:268. 1806.
Scirpus lacustris L. var. *condensatus* Peck, Ann. Rep. N.Y. State Mus. 53:853. 1900.
Scirpus validus Vahl var. *creber* Fern. Rhodora 45:282. 1943.

Scirpus validus Vahl. var. *creber* Fern. f. *megastachys* Fern. Rhodora 45:282. 1943.
Scirpus validus Vahl. var. *condensatus* (Peck) Beetle, Univ. Wyom. Publ. 13:6. 1948.
Scirpus validus Vahl. var. *condensatus* (Peck) Beetle f. *megastachys* (Fern.) Beetle, Univ. Wyom. Publ. 13:6. 1948.
Schoenoplectus validus (Vahl) A. Love & D. Love, Bull. Torrey Club 81:33. 1954.
Scirpus lacustris L. ssp. *validus* (Vahl) Koyama, Can. Journ. Bot. 40:927. 1962.

Coarse perennial from stout rhizomes; culms erect, robust, terete, to 2.7 m tall, soft, glabrous; sheaths usually without blades; involucral leaf 1, appearing like the continuation of the culm, up to 3 cm long; panicle much branched, the rays pubescent; spikelets ovoid, subacute, 5–9 (–15) mm long; scales ovate, mucronulate, equaling to longer than the achene, pubescent at tip and along midvein; bristles stout, retrorsely hairy, as long as to longer than the achene; achenes obovoid, plano-convex, 1.5–2.3 (–2.8) mm long, 1.3–1.6 (–1.8) mm broad, brown. May–September.

Swamps, marshes, sloughs, ditches, around ponds and lakes, along rivers, often in standing water.

IA, IL, IN, KS, KY, MO, NE, OH (OBL). The U.S. Fish and Wildlife Service calls this species *Scirpus tabernaemontani*.

Softstem bulsedge.

This very common species has very soft culms that crush easily due to the presence of large air cells in the culm.

16. **Schoenoplectus torreyi** (Olney) Palla, Allg. Bot. Zeit. Syst. 17:Beibl. 3. 1911. Fig. 170.
Scirpus torreyi Olney, Proc. Prov. Frankl. Soc. 1:32. 1847.
Scirpus subterminalis Torr. var. *cylindricum* (Torr.) Koyama, Can. Journ. Bot. 40:930. 1962.

Perennial from weak rhizomes; culms erect, solitary, 3-angled, to 1.5 m tall, glabrous; leaves 2–3 near the base, obtuse, otherwise reduced to bladeless sheaths; involucral leaf 1, appearing as a continuation of the culm, erect, obtuse, to 15 cm long; inflorescence a small cluster of 1–4 lateral, sessile spikelets; spikelets ovoid-cylindric, subacute, to 15 mm long; scales ovate, mucronate, stramineous to pale brown; bristles pale, longer than the achene; achenes obovoid, long-beaked, trigonous, 3–4mm long, brown. June–September.

Around ponds, swamps, in deep water of sinkhole ponds.

IA, IL, IN, MO, NE, OH (OBL). This species is called *Scirpus torreyi* by the U.S. Fish and Wildlife Service.

Torrey's bulsedge.

The long-beaked achenes readily distinguish this species from all others in the genus.

15. **Scirpus** L.—Bulsedge

Perennials with rhizomes; stems unbranched, terete or 3-angled, glabrous; leaves basal and alternate on the culm, flat, minutely toothed; inflorescence terminal, in irregular panicles or umbels, subtended by 2–several bracts; spikelets numerous, pedicellate or sessile, many-flowered; scales glabrous; bristles 5–6, or 0; stamens 3; style 2- or 3-cleft; achenes trigonous or lenticular.

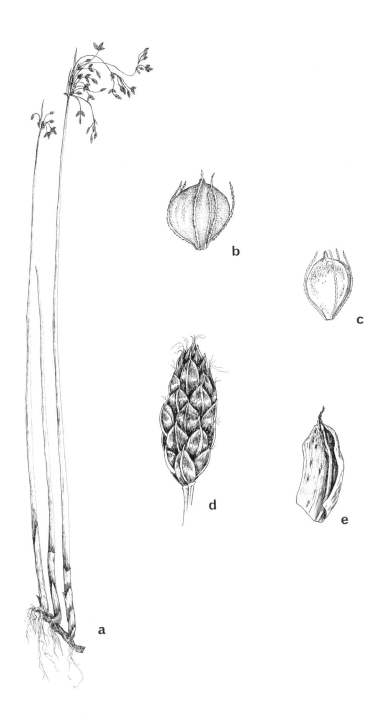

169. *Schoenoplectus*
tabernaemontani.

a. Habit.
b, c. Achenes.

d. Spikelet.
e. Scale.

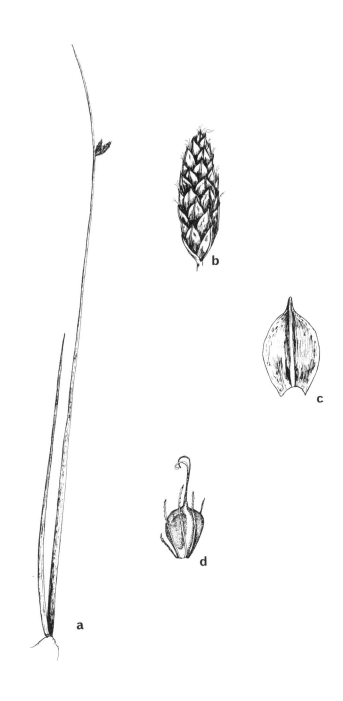

170. *Schoenoplectus torreyi.* a. Habit.
                                        b. Spikelet.   c. Scale.
                                                                         d. Achene.

With the removal of several species to other genera such as *Bolboschoenus, Schoenoplectus,* and *Trichophorum, Scirpus* consists of about 30 species distributed worldwide.

1. Leaves 10–20.
    2. Spikelets up to 4 mm long, sessile in small clusters; bristles longer than the achene ....
    .... 12. *S. polyphyllus*
    2. Spikelets 4–15 mm long, most or all of them pedicellate; bristles shorter than to about as long as the achene .... 4. *S. divaricatus*
1. Leaves 2–10.
    3. Bristles smooth, usually twisted and conspicuously longer than the achenes, particularly at maturity (sometimes not conspicuous in *S. pendulus*).
        4. Bristles at maturity as long as to twice as long as the achenes; scales usually with a prominent green midvein .... 11. *S. pendulus*
        4. Bristles at maturity many times longer than the achenes; scales usually with an inconspicuous midvein.
            5. Spikelets and involucels reddish brown .... 3. *S. cyperinus*
            5. Spikelets and involucels whitish brown or blackish.
                6. Spikelets and involucels whitish brown .... 10. *S. pedicellatus*
                6. Spikelets and involucels blackish .... 1. *S. atrocinctus*
    3. Bristles barbed, not twisted, shorter than to 1 1/2 times longer than the achenes at maturity, or bristles absent.
        7. Styles 2-cleft; achenes lenticular .... 8. *S. microcarpus*
        7. Styles 3-cleft; achenes trigonous.
            8. Bristles 0–3, shorter than the achenes .... 6. *S. georgianus*
            8. Bristles usually 5 or 6, up to 1 1/2 times longer than the achenes.
                9. Lowest sheaths reddish to red-purple .... 5. *S. expansus*
                9. Lowest sheaths greenish or brown.
                    10. Lower leaf blades and sheaths septate (with conspicuous cross-walls).
                        11. Scales with an awn 0.4–0.8 mm long .... 9. *S. pallidus*
                        11. Scales with an awn less than 0.4 mm long .... 2. *S. atrovirens*
                    10. Lower leaf blades and sheaths not septate .... 7. *S. hattorianus*

1. **Scirpus atrocinctus** Fern. Proc. Am. Acad. 34:502. 1899. Fig. 171.

Perennial from slender rhizomes; culms erect, very slender, to 1.5 m tall, glabrous; leaves to 5 mm broad; involucral leaves 3–5, exceeding the umbel; involucels black; spikelets numerous, the lateral pedicellate, black, ovoid, 3–6 mm long; scales ovate-lanceolate, subacute, 1.4–2.0 mm long; bristles many times longer than the achenes; achenes lanceoloid, greenish, apiculate, 0.7–1.0 mm long.

Swamps, bogs.

IA, IL (OBL).

Black bulsedge.

This species has the general appearance of *S. cyperinus* but differs by its conspicuous blackish involucels, spikelets, and scales.

2. **Scirpus atrovirens** Willd. Enum. Pl. 79. 1809. Fig. 172.
Scirpus atrovirens Willd. f. *proliferus* F. J. Hermann, Rhodora 40:77. 1938.

Tufted perennial from short rhizomes; culms erect, 1–several, to about 1.5 m tall, glabrous; leaves to 20 mm broad, usually nodulose-septate, as are the sheaths;

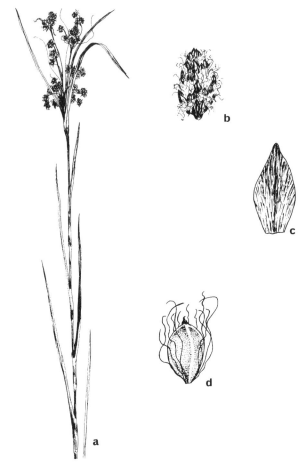

171. *Scirpus atrocinctus.*   a. Habit.   c. Scale.
                             b. Spikelet.   d. Achene.

inflorescence paniculate, with some axillary bulblets often present at maturity; spikelets usually brownish, ovoid, to 6 mm long, in numerous glomerules, becoming blackish green at maturity; scales 1.4–2.1 mm long, elliptic, mucronate; bristles 6, shorter than to slightly exceeding the achenes, with retrorse, round-tipped teeth in the upper two-thirds; achenes 1.0–1.3 mm long, ellipsoid or obovoid.

Sloughs, ditches, around lakes and ponds, along rivers and streams, sometimes in shallow water.

IA, IL, IN, KS, KY, MO, NE, OH (OBL).

Dark green bulsedge.

This is a very widespread, common species that is found in almost every type of wetland. It differs from the very similar *S. pallidus* by lacking awn-tipped scales, although a minute cusp may be present in *S. atrovirens.* It is not uncommon to find plantlets growing from bulblets in the inflorescence of this species.

172. *Scirpus atrovirens.*
a. Habit.

b. Proliferating inflorescence.
c. Spikelet.

d. Scale.
e. Achene.

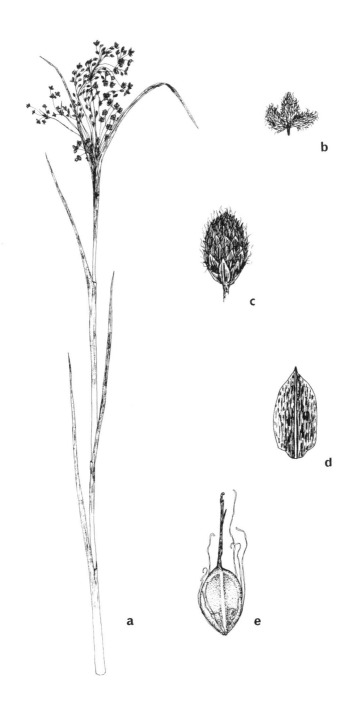

173. *Scirpus cyperinus.*
a. Habit.

b. Spikelets.
c. Spikelet

d. Scale.
e. Achene.

3. **Scirpus cyperinus** (L.) Kunth, Enum. Pl. 2:170. 1837. Fig. 173.
Eriophorum cyperinum L. Sp.Pl. ed. 2, 1:77. 1762.
Scirpus eriophorum Michx. Fl. Bor. Am. 1:33. 1803.
Scirpus eriophorum Michx. var. *cyperinus* (L.) Gray, Man. Bot., e. 2, 501. 1856.
Scirpus cyperinus (L.) Kunth var. *eriophorum* (Michx.) Kuntze, Rev. Gen. Pl. 2:757. 1891.
Scirpus cyperinus (L.) Kunth var. *pelius* Fern. Rhodora 8:164. 1906.
Scirpus rubricosus Fern. Rhodora 47:124. 1945.
Scirpus cyperinus (L.) Kunth var. *rubricosus* (Fern.) Gilly, Iowa State Coll. Journ. Sci. 21:82. 1946.

Very leafy, tufted perennial from stout rhizomes; culms erect, terete or somewhat angular, to 2 m tall; leaves crowded near the base of the plant, linear; involucral bracts 3–6, exceeding the umbel; involucels reddish brown or rarely drab; spikelets numerous, all sessile or subsessile or the lateral spikelets pedicellate, reddish brown, rarely drab, ovoid, woolly when mature, 3–6 mm long; scales ovate-lanceolate, obtuse to subacute, reddish brown; bristles reddish brown, rarely drab, much exceeding the scales; achenes pale, apiculate, lanceoloid, 0.7–1.0 mm long. July–October.

Swamps, marshes, ditches, sloughs, fens, around lakes and ponds.

IA, IL, IN, KS, MO, NE (OBL), KY, OH (FACW+).

Wool grass.

This large, clump-forming sedge is distinguished by its reddish involucels and long, reddish bristles that exceed the spikelets when mature, creating a fuzzy appearance to the inflorescence. This is a variable species. Plants in which the lateral spikelets are pedicellate may be designated as var. *rubricosus*. Plants with drab-colored scales and bristles may be called var. *pelius*.

Scirpus cyperinus closely resembles *S. pedicellatus* and *S. atrocinctus*. *Scirpus cyperinus* has all its spikelets sessile or subsessile or the lateral spikelets occasionally pedicellate. The spikelets and involucels are usually reddish brown, although rarely they may be drab. *Scirpus pedicellatus* has lateral spikelets all pedicellate, but the spikelets, bristles, and involucels are pale brown to whitish. *Scirpus atrocinctus* has the lateral spikelets all pedicellate, but the spikelets are blackish as are the involucels.

4. **Scirpus divaricatus** Ell. Sketch Bot. S. Carol. 1:88. 1816. Fig. 174.
Isolepis divaricatus (Ell.) A. Dietr. Sp.Pl., ed. 6, 2:131. 1833.

Tufted perennial; culms 3-angled, ascending to spreading, to 1.5 m tall, glabrous; leaves 10–20, alternate on the culm, 4–10 mm broad, serrate; inflorescence paniculate, spreading to pendulous, each branchlet bearing one sessile spikelet and several pedicellate spikelets, subtended by 3–6 bracts; spikelets

174. *Scirpus divaricatus.* Habit, spikelet, achene, and scale.

ovoid, acute at tip, up to 15 mm long; scales ovate to elliptic, acute to subacute at tip, 1.2–1.8 mm long, brown or reddish brown; bristles 6, smooth, shorter than to about as long as the achene; styles 3-cleft; achenes trigonous, obovoid to ellipsoid, 0.7–1.0 mm long, smooth or minutely pebbled, yellow to brown. June–September.

Swamps, ditches, sometimes in shallow water.

MO (OBL).

Spreading bulsedge.

The numerous leaves and spreading to pendulous inflorescence readily distinguish this southern species.

5. **Scirpus expansus** Fern. Rhodora 45:293–295, pl. 767, f. 1–3. 1943. Fig. 175.

Perennial from reddish rhizomes; stems erect, 3-angled, stout, to 2 m tall, glabrous; sheaths at base of plant reddish; leaves 1–2 cm wide; inflorescence much branched, bearing numerous spikelets in sessile clusters; spikelets ovoid to cylindric, 2.5–5.0 mm long; scales ovate, obtuse, mucronulate, 1.5–2.2 mm long, the margins blackish; bristles 6, shorter than to a little longer than the achene; styles 3-cleft; achenes trigonous, obovoid, 1.0–1.5 mm long, short-beaked, purple-brown to brown. July–August.

Swamps, along streams.

OH (OBL).

Robust bulsedge.

This very stout species is conspicuous by virtue of its lower reddish sheaths.

175. *Scirpus expansus.* Habit, stigmas, achene, and spikelets.

6. **Scirpus georgianus** Harper, Bull. Torrey Club 27:331. 1900. Fig. 176.
Scirpus atrovirens Willd. var. *georgianus* (Harper) Fern. Rhodora 23:134. 1921.

Tufted perennial from short rhizomes; culms erect, 1–several, to about 1.5 m tall, glabrous; leaves to nearly 20 mm broad, nearly smooth or sometimes nodulose-septate, as are the sheaths; inflorescence paniculate, with some axillary bulblets often present at maturity; spikelets brownish or rarely blackish, ovoid, to 4 mm long, in numerous glomerules; scales 1.2–1.8 mm long, elliptic, mucronulate; bristles absent or 1–3, shorter than the achene, with a few teeth near the tip; achenes trigonous, 0.8–1.1 mm long, ellipsoid or obovoid. May–September.

Ditches, sloughs, along rivers and streams, around lakes and ponds, sometimes in shallow water.

IA, IL, IN, KS, KY, MO, NE, OH (OBL).

Common bulsedge.

176. *Scirpus georgianus.*
a. Habit.
b. Spikelet.
c. Scale.
d. Achene.

Although this species looks very much like *S. atrovirens* and for many years was not distinguished from it, it differs by its fewer bristles and slightly smaller spikelets and achenes.

7. **Scirpus hattorianus** Mak, Journ. Jap. Bot. 8:441. 1933. Fig. 177.

Tufted perennial from short rhizomes; culms erect, 1–several, to about 1.5 m tall, glabrous; leaves nearly to 20 mm broad, scarcely or not at all nodulose-septate; inflorescence paniculate, with axillary bulblets sometimes present at maturity; spikelets blackish or dark brown, ovoid, to 3.5 mm long, in numerous glomerules; scales 1.2–2.0 mm long, elliptic, mucronate; bristles 5 or 6, shorter than to about equaling the achene, with retrorse round-tipped teeth in the upper half; achenes 0.8–1.1 mm long, ellipsoid or obovoid. July–September.
Ditches.
IL, IN (OBL).
Smooth bulsedge.
This species has the general appearance of *S. atrovirens,* but it lacks the septations on the leaves and sheaths and has darker scales and usually shorter bristles.

8. **Scirpus microcarpus** Presl, Rel. Haenk. 1:195. 1828. Fig. 178.
Scirpus rubrotinctus Fern. Rhodora 2:20. 1900.

Perennial from a rather thick reddish rhizome; culms erect, stout, nearly 1 m tall, glabrous; sheaths reddish at base; leaves to 15 mm broad, faintly nodulose-septate; involucral leaves 3, the longest as long as or longer than the inflorescence; panicle rather stiff, much branched; spikelets narrowly ovoid, 3–8 (–13) mm long, in several glomerules; scales 1.1–3.1 mm long, ovate to elliptic, obtuse, mucronulate, green and black; bristles 4, hairy throughout, shorter than to about 1 1/2 times longer than the achene, with retrorse, sharp-pointed teeth nearly to the base; achenes 0.7–1.6 mm long, lenticular, whitish.
July–September.
Marshes, swamps, occasionally in shallow water.
IA, IL, NE (OBL).
Small-fruited bulsedge.
The lowermost sheaths are red-tinged and the bristles are hairy their entire length.

9. **Scirpus pallidus** (Britt.) Fern. Rhodora 8:163. 1906. Fig. 179.
Scirpus pallidus Willd. var. *pallidus* Britt. Trans. N.Y. Acad. Sci. 9:14. 1889.

Perennial with short rhizomes; culms erect, 3-angled only in the upper part, to 1.5 m tall, glabrous; leaves several, but less than 10, nodulose-septate, up to 20 mm broad, serrulate; inflorescence paniculate, each branch terminating in a cluster of sessile spikelets; spikelets ovoid, obtuse to acute at tip, 3.5–8.0 mm long; scales obovate to elliptic, obtuse at tip but short-awned, 1.8–2.8 mm long, pale brown but becoming black at maturity; bristles 5–6, up to 1 1/2 times longer than the achene; styles 3-cleft; achenes trigonous, ellipsoid, 1.0–1.3 mm long, smooth or faintly pebbled, pale brown. May–September.

177. *Scirpus hattorianus.*   a. Habit.   c. Scale.
                                   b. Spikelet.   d. Achene.

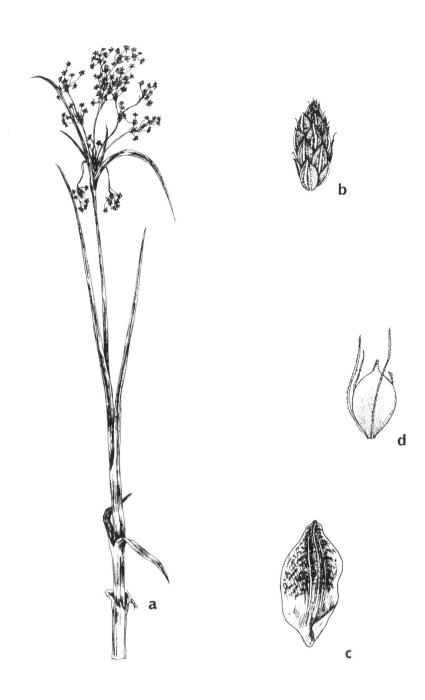

178. *Scirpus microcarpus.*

a. Habit.
b. Spikelet.
c. Scale.
d. Achene.

Wet depressions, often in standing water.

IA, KS, MO, NE (OBL).

Pale bulsedge.

This primarily western species is sometimes considered to be a variety of *S. atrovirens* from which it differs by its pale scales and pale achenes, although the scales turn black at maturity.

10. **Scirpus pedicellatus** Fern. Rhodora 2:16. 1900. Fig. 180.
Scirpus cyperinus (L.) Kunth var. *pedicellatus* (Fern.) Schuyl. Mich. Bot. 1:80. 1962.

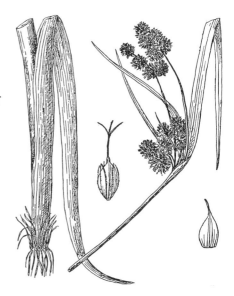

179. *Scirpus pallidus.* Habit, achene, and scale.

Coarse, tufted perennial from short rhizomes; culms erect, rather stout, terete, to about 1.5 m tall; leaves narrow, to 10 mm broad; involucral leaves 4–7, exceeding the umbel; involucels pale brown to whitish; spikelets numerous, fascicled, the central spikelet sessile, the lateral spikelets pedicellate, pale brown to whitish, ovoid, woolly when mature, 3–6 mm long; scales 1.4–1.8 mm long, ovate-lanceolate, obtuse to sub-acute, whitish brown; bristles pale brown to whitish, exceeding the scales; achenes very pale, apiculate, lanceoloid, 0.7–1.0 mm long. May–July.

Marshes, around lakes and ponds, along rivers and streams.

IL, MO, OH (OBL).

Stalked bulsedge.

This species resembles *S. cyperinus* but differs by its whitish brown spikelets and involucels.

11. **Scirpus pendulus** Muhl. Cat. 7. 1813. Fig. 181.

Perennial from short rhizomes; culms erect, rather slender but firm, to 1.5 m tall, glabrous; leaves 5–10, widely separated on the culm, to 8 mm broad; involucral leaves several, brown; inflorescence umbellate, 15–20 cm long, the rays slender and pendulous; spikelets oblong-cylindric, to 10 mm long, 2–4 mm broad; scales ovate, acute, pale brown to castaneous with a conspicuous green midvein, 1.7–2.2 mm long; bristles pale, glabrous or nearly so, longer than the achene; achenes biconvex to obscurely trigonous, ellipsoid, apiculate, papillate, 0.8–1.3 mm long, pale brown to whitish. May–August

Fens, ditches, along rivers and streams, around ponds and lakes, marshes, springs.

IA, IL, IN, KS, KY, MO, NE, OH (OBL).

Nodding bulsedge.

This species is recognized by its pendulous inflorescences and the scales of the spikelets that have a conspicuous midnerve.

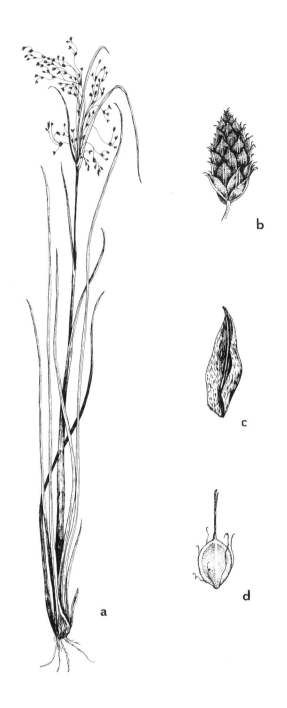

180. *Scirpus pedicellatus.* a. Habit. c. Scale.
b. Spikelet. d. Achene.

181. *Scirpus pendulus.*   a. Habit.   c. Scale.
                                     b. Spikelet.   d. Achene.

12. **Scirpus polyphyllus** Vahl, Enum. Pl. 2:274. 1806. Fig. 182.

Perennial from fibrous rhizomes; culms erect, rather slender, to 1.5 m tall, glabrous; leaves 10–20, distichous, crowded, long-attenuate, to 10 mm broad; involucral bracts 2–5, usually shorter than the inflorescence; spikelets in numerous small glomerules, ovoid, subacute, pedicellate, 2.5–4.0 mm long, reddish brown; scales suborbicular, obtuse, mucronate, 1.0–1.5 mm long; bristles 6, about twice as long as the achene; achenes oblanceoloid to narrowly obovoid, apiculate, 0.9–1.3 mm long, pale. June–September.

Seeps, low woods, fens, along streams.

IL, IN, MO, OH (OBL).

Many-leaved bulsedge.

This species may have as many as 20 leaves along the culm. The short spikelets never more than 4 mm long distinguish this species from *S. divaricatus*, the only other species of *Scirpus* in the central Midwest with more than 10 leaves on the culm.

15. **Trichophorum** Pers.

Tufted annuals or perennials; culms unbranched, terete or 3-angled, glabrous; lowermost leaves reduced to bladeless sheaths; inflorescence terminal, consisting of one spikelet subtended by an involucral bract; spikelets 2- to 8-flowered; at least some of the scales awned; bristles several; stamens 3; style 3-cleft; achenes trigonous.

The eight species that comprise this genus are often placed in the genus *Scirpus*.

These plants differ from *Scirpus* by their solitary, terminal spikelet on each culm.

1. **Trichophorum cespitosum** (L.) Schur var. **callosum** (Bigel.) Mohlenbr. Ill. Fl. Il., Sedges, ed. 2, 209. 2001. Fig. 183.
Scirpus cespitosus L. var. *callosus* Bigel, Fl. Bost., ed. 2, 21. 1824.

Densely cespitose annual, without stolons; culms very wiry, filiform, terete, to 70 cm tall, with numerous coriaceous sheaths at the base; upper sheaths blade-bearing; blade terete to involute, callus-tipped, to 1.5 cm long; spikelets ovoid-lanceoloid, 3–6 mm long; scales pale brown, the lowest awned, the uppermost obtuse; bristles slender, without barbs, nearly twice as long as the achene; achenes obovoid, trigonous, plano-convex, 1.5–2.0 mm long. July–September.

Bogs.

IL (OBL). The U.S. Fish and Wildlife Service calls this species *Scirpus cespitosus*.

Tufted bulsedge.

This is primarily a species of the tundra that barely reaches the central Midwest. The single spikelet at the tip of each culm and the short, nearly basal leaves, are distinctive for this species.

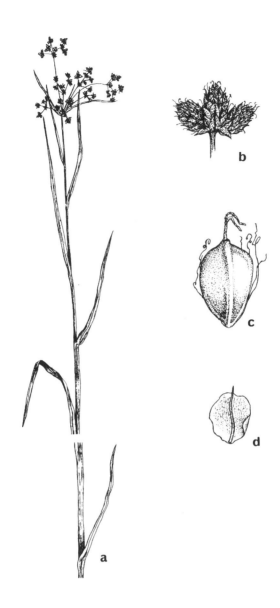

182. *Scirpus polyphyllus.*   a. Habit.   c. Achene.
                                         b. Spikelets.   d. Scale.

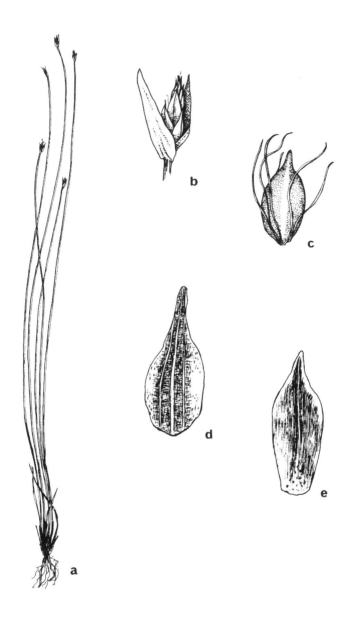

183. *Trichophorum cespitosum* var. *callosum.*

a. Habit.
b. Spikelet.

c. Achene.
d, e. Scales.

Glossary

achene. A type of 1-seeded, dry, indehiscent fruit with the seed coat not attached to the mature ovary wall.

acicular. Needlelike.

acuminate. Gradually tapering to a long point.

acute. Sharply tapering to a short point.

androgynous. With staminate flowers above, pistillate flowers below.

annual. Living only for one year.

anther. The terminal part of a stamen that bears the pollen.

antrorse. Projecting forward.

apiculate. Abruptly short-pointed at the tip.

appressed. Lying flat against a surface.

approximate. Crowded or near to each other, often touching; opposed to remote.

aristate. Bearing a short awn.

ascending. Pointing upward.

attenuate. Gradually becoming narrowed.

awn. A bristle terminating a structure.

awned. Bearing an awn.

awnless. Without an awn.

basifixed. Said of an anther attached at the base.

beak. The narrow terminal tip of some perigynia.

biconvex. Rounded on both the front and back faces.

bidentate. With two teeth.

bifid. Two-cleft.

bract. An accessory structure at the base of a flower, sometimes appearing leaflike, other times setaceous or scalelike.

bristle. A stiff hair or hairlike growth; a seta.

bulblet. A small bulb, or mass of tissue capable of developing into a new plant.

canaliculate. Grooved.

capillary. Threadlike.

capitate. Forming a head.

cartilaginous. Firm but flexible.

castaneous. Chestnut colored.

cauline. Belonging to a stem or culm.

cespitose. Growing in tufts.

ciliate. Bearing short, marginal hairs.

clavate. Club-shaped.

compressed. Flattened.

concave. Curved on the inner surface; opposite of convex.

conduplicate. Folded together lengthwise.

confluent. Running together.

conic. Cone-shaped.

conical. Cone-shaped.

contiguous. Touching the next one in a series.

convex. Curved on the outer surface; opposite of concave.

convolute. Rolled lengthwise.

coriaceous. Leathery.

corm. An underground, vertical stem with scaly leaves, differing from a bulb by lacking fleshy leaves.

corrugate. Wrinkled or in folds.

cucullate. Hooded.

culm. The stem that terminates in an inflorescence.

cuneate. Wedge-shaped, tapering to the base.

cusp. A very short point.

cuspidate. Terminating in a very short point.

cylindric. Shaped like a cylinder.

cylindrical. Shaped like a cylinder.

cyme. A type of broad and flattened inflorescence in which the central flowers bloom first.

cymose. Bearing a cyme.

deciduous. Falling away.

decumbent. Lying flat, but with the tip ascending.

decurrent. Extending beyond the point of attachment.

deltoid. Referring to a solid object that is triangular.

dentate. With sharp teeth, the tips of which project outward.

depressed. Shallowly sunken in.

dilated. Expanded, swollen.

dioecious. Having staminate and pistillate flowers on separate plants.

disarticulating. Breaking off.

distended. Swollen over a structure.

distichous. Two-ranked; arranged in two vertical rows.

divaricate. Widely spreading.

divergent. Spreading.

dorsal. The outer face of a structure.

ebarbellate. Without barbs.

ellipsoid. Referring to a solid object that is broadest at the middle, gradually tapering to both ends; narrower than oblongoid.

elliptic. Broadest at the middle, gradually tapering to both ends; narrower than oblong.

emarginate. Having a shallow notch at the tip.

excurrent. Extending beyond the margin of a structure from which it originates.

fascicled. Clustered.

ferruginous. Iron-colored.

fibrillose. Bearing numerous fine fibers.

fibrous. Referring to roots borne in tufts.

filament. That part of the stamen supporting the anther.

filiform. Threadlike.

flaccid. Weak, flabby.

flexuous. Flexible; zigzag.

foliaceous. Leaflike.

fusiform. Spindle-shaped.

gibbous. Swollen on one side.

glabrous. Without pubescence or hairs.

glaucous. With a whitish covering that can be wiped off.

globoid. Referring to a round, solid body.

globose. Round; globular.

glomerulate. Bearing small compact clusters.

glomerule. A small compact cluster.

glume. A scale, usually sterile, in a spikelet.

granular. Having the surface texture of tiny grains.

gynecandrous. A spike with pistillate flowers at the tip and staminate flowers below.

hirtellous. With minute stiff hairs.

hispidulous. With minute stiff hairs.

hyaline. Transparent.

hypogynium. A structure that subtends the ovary.

imbricate. Lapped over each other.

impressed. Sunken into the surface.

inflorescence. A cluster of flowers or spikes.

intervenal. Between the veins.

invaginated. Notched; sunken inward.

involucel. A secondary involucre.

involucral. Referring to a circle of bracts that subtends a flower cluster.

involucre. A circle of bracts that subtends a flower cluster.

involute. Rolled inward.

keel. A central ridge.

keeled. Bearing a ridgelike process.

lanceolate. Lance-shaped; broadest near the base, gradually tapering to the narrower apex; narrower than ovate.

lanceoloid. Referring to a solid object that is broadest near the base, gradually tapering to the narrower apex; narrower than ovoid.

lenticular. Lens-shaped.

ligule. The structure at the summit of the ventral side of the sheath where it merges into the blade.

linear. Elongated and uniform in width throughout.

locular. Referring to the cells of a compound ovary.

locule. A cell or cavity of a compound ovary.

lustrous. Shiny.

maculate. Spotted.

margin. The border of something.

membranous. Membranelike; very thin, often translucent.

mitriform. Cap-shaped.

moniliform. Beadlike.

mucro. A short, terminal point.

mucronate. Possessing a short, abrupt tip.

node. The place on the stem or culm from which leaves and spikes arise.

nodulose. Bearing small knots or knobs.

obconic. Reversely cone-shaped.

oblanceoloid. Referring to a solid object that is broadest at the apex, tapering gradually to the narrow base.

oblique. At an angle.

oblong. Broadest at the middle, tapering to both ends; broader than elliptic.

oblongoid. Referring to a solid object that is broadest at the middle, tapering to both ends; broader than ellipsoid.

obovate. Broadly rounded at the apex, becoming narrowed below; broader than oblanceolate.

obovoid. Referring to a solid object that is broadly rounded at the apex, becoming narrowed below.

obpyramidal. Reverse pyramid-shaped.

obtrulloid. Reversely shaped like a brick-layer's trowel.

obtuse. Rounded; blunt.

opaque. Referring to an object that cannot be seen through.

orbicular. Round.

oval. Broadly elliptic.

ovary. The lower swollen part of the pistil that produces the ovules.

ovate. Broadly rounded at the base, becoming narrowed above; broader than lanceolate.

ovoid. Referring to a solid object that is broadly rounded at the base, becoming narrowed above; broader than lanceoloid.

ovule. The egg-producing structure found within the ovary.

panicle. A type of inflorescence composed of several racemes.

paniculate. In the form of a panicle.

papilla. A small, pimplelike projection.

papillate. Bearing small, pimplelike projections.

papillose. Warty.

peduncle. The stalk of an inflorescence or of a spike.

pedunculate. Bearing a peduncle.

pellucid. Clear; transparent.

pendulous. Drooping; nodding.

perigynium. A saclike covering enclosing the achene in Carex.

perennial. Living more than two years.

perfect. Bearing both fertile stamens and pistils in the same flower.

perianth. Those parts of a flower including both the calyx and corolla.

pilose. Bearing soft hairs.

pistil. The female reproductive organ.

pistillate. Bearing pistils but not stamens.

plano-convex. Flat on one surface, rounded on the other.

prostrate. Lying flat on the substrate.

puberulent. With minute hairs.

pubescent. Bearing some kind of hairs.

punctate. Dotted.

puncticulate. With small dots.

pyramidal. Pyramid-shaped.

pyriform. Pear-shaped.

rachilla. A small rachis; the axis to which the parts of a spikelet are attached.

rachis. The primary axis of a structure.

rank. Referring to the number of planes in which structures are borne.

ray. A branch of an inflorescence.

reclining. Lying down.

recurved. Turned backward or downward.

reflexed. Turned downward.

remote. Apart; referring to spikes that do not touch each other.

reticulate. Resembling a network.

retrorse. Pointing downward.

retuse. Shallowly notched at a rounded apex.

revolute. Rolled under from the margins.

rhizomatous. Bearing rhizomes.

rhizome. An underground horizontal stem, bearing nodes, buds, and roots.

rhombic. Quadrangular; four-sided.

ribbed. With elevated veins.

rootstock. The underground root system; sometimes synonymous with rhizome.

rotund. Rounded in outline.

rudimentary. Poorly developed.

rugulose. Wrinkled-appearing.

russet. Red-brown.

scaberulous. Slightly roughened to the touch.

scabrous. Rough to the touch.

scale. A modified leaf that subtends each staminate and each pistillate flower; a structure that is formed on rhizomes.

scarious. Thin and membranous.

septate. With dividing walls.

serrate. With teeth that project forward.

serrulate. With very small teeth that project forward.

sessile. Without a stalk.

setaceous. Bristlelike.

sheath. A tubular part of the leaf that surrounds the culm.

spatulate. Oblong, but with the basal end elongated.

spike. The basic unit of the inflorescence in Carex.

spikelet. The basic unit in a sedge inflorescence.

squarrose. Referring to the tips of perigynia that bend outward or downward.

stamen. The male reproductive organ.

staminate. Bearing stamens but not pistils.

stigma. The terminal part of the pistil that receives pollen.

stipitate. Possessing a short stalk.

stolon. A slender, horizontal stem on the surface of the ground.

stoloniferous. Bearing stolons.

stramineous. Straw-colored.

striate. Marked with grooves.

style. That part of the pistil between the ovary and the stigmas.

subacute. Nearly tapering to a short point.

subaristate. With an extremely short awn.

subcanaliculate. Nearly forming a canal.

subcoriaceous. Almost leathery.

subcylindric. Nearly cylindric.

subglobose. Almost globe-shaped.

submembranous. Nearly membranelike or thin.

suborbicular. Nearly round.

subsessile. With a very short stalk.

substipitate. Possessing a very short stalk.

subterete. Nearly round in cross-section.

subulate. With a very short, narrow point.

subuloid. Referring to a solid object that is long-tapering to the apex.

suffused. Spread throughout; flushed.

terete. Round in cross-section.

trabecula. A cross-marking.

translucent. Allowing some light to pass through; intermediate between opaque and transparent.

transverse. Lying across.

trigonous. Triangular in cross-section.

trulloid. Shaped like a bricklayer's trowel.

truncate. Abruptly cut across.

tuber. An underground fleshy stem formed as a storage organ at the end of a rhizome.

tubercle. A small, usually rounded projection.

tubular. Forming a tube.

turgid. Swollen to the point of bursting.

tussock. A clump.

umbel. A type of inflorescence in which the flower stalks arise from the same level.

umbelliform. Having the form of an umbel.

umbonate. Having a protuberance or rounded elevation.

unisexual. Bearing either staminate or pistillate flowers, but not both.

ventral. The inner face of a structure.

verrucose. Warty.

versatile. Said of an anther attached at the middle.

villous. With long, soft, slender, unmatted hairs.

viscid. Sticky.

Illustration Credits

Illustrations 1–17, 29, 30, 33, 35–37, 39–52, 55–58, 60, 66–71, 73–75, 77–94, 96, 98–102, 104–9, 111, 115–18, 120–22, 124–26, 128, 132–35, 137, 139, 142–44, 147–52, 154, 155, 157, 158, 160–62, 164, 165, 167–73, 176–78, 180–83 were prepared by Paul W. Nelson. Copyright 2005 by Paul W. Nelson.

Illustrations 28, 31, 32, 34, 38, 53, 54, 59, 61, 63, 65, 72, 89, 95, 103, 110, 112, 123, 129, 130, 131, 138, 140, 145, 146, 156, 159, 174, 175, 179 are reprinted from *An Illustrated Flora of the Northern United States and Canada,* by Nathaniel Britton and Addison Brown. Charles Scribner's Sons, 1913; reprinted 1970 by Dover Publications.

Illustrations 62, 76, and 153 are reprinted from *Steyermark's Flora of Missouri,* by George Yatskievych. Missouri Botanical Garden Press, 1999. Illustrations 62 and 76 are by Paul Nelson; illustration 153 is by Phyllis Bick. Reprinted by permission of Missouri Botanical Garden Press.

Illustrations 97, 113, 114, and 141 are reprinted from *Aquatic and Wetland Plants of the Southeastern United States: Monocotyledons,* by Robert K. Godfrey and Jean W. Wooten. Copyright 1979 by the University of Georgia Press. Reprinted by permission of the University of Georgia Press.

Illustration 136 is reprinted from *Sida, Contributions to Botany* 4.2 (1971): 168. Reprinted by permission of the Botanical Research Institute of Texas and artist Robert Kral.

Illustrations 127 and 163 are reprinted from *Vascular Plants of the Pacific Northwest,* by C. L. Hitchcock et al. University of Washington Press, 1977. Drawn by Jeanne R. Janish. Reprinted by permission of the University of Washington Press.

Illustration 166 is reprinted from *Shinner's and Mahler's Flora of North Central Texas,* by George M. Diggs Jr., Barney L. Lipscomb, and Robert O. Kennon. Botanical Research Institute of Texas, 1999. Reprinted by permission of Botanical Research Institute of Texas and artist Linny Heagy.

Index to Genera and Species

Names in roman type are accepted names, while those in italics are synonyms and are not considered valid. Page numbers in bold refer to pages that have illustrations.

atrocinctus 243, **244**
atropurpureus 159
atrovirens 243, **245**
autumnalis 198
baldwinianus 196
californicus 224
caribaeus 164
cespitosus 256
cyperinus **246**, 247
debilis 233
deltarum 226
divaricatus **247**
equisetoides 164
eriophorum 247
etuberculatus 227
expansus **248**
fluviatilis 2
geniculatus 164
georgianus 248, **249**
glaucus 164, 221
hallii 228
hattorianus 250, **251**
heterochaetus 229
intermedius 168
juncoides 233
lacustris 221, 239
malhuerensis 221
maritimus 2
micranthus 204, 205
microcarpus 250, **252**
mucronatus 198
mucronulatus 198
nanus 180
nevadensis 231
nitens 208
obtuse 175
occidentalis 221
olivaceus 172
olneyi 224
ovatus 174
pallidus 250, **253**
paludosus 2
palustris 175
parvulus 180
pauciflorus 181
pedicellatus 253, **254**
pendulus 253, **255**
polyphyllus 256, **257**
pungens 233
purshianus 233
quadrangulatus 181

quinqueflora 181
rostellatus 184
rubricosus 247
rubrotinctus 250
saximontanus 233
smithii 237
subterminalis 237
supinus 228, 233
tabernaemontani 239
torreyi 240
uninodis 228
vahlii 199
validus 239
wolfii 186

Trachelostylis 198
  borealis 198
Trichophorum 256
  cespitosum 256, **258**
Trichophyllum 172
  olivaceum 172
  ovatum 174

Index to Common Names

Robert H. Mohlenbrock taught botany at Southern Illinois University at Carbondale for thirty-four years, obtaining the title of Distinguished Professor. Since his retirement in 1990, he has served as senior scientist for Biotic Consultants Inc., teaching wetland identification classes around the country. Mohlenbrock has been named SIU Outstanding Scholar and has received the SIU Alumnus Teacher of the Year Award, the College of Science Outstanding Teacher Award, and the Meritorious Teacher of the Year Award from the Association of Southeastern Biologists. Since 1984, he has been a monthly columnist for *Natural History* magazine. He is the author of 50 books and more than 560 publications.